Learn, Practice, Succeed

Eureka Math®
Grade 6
Module 5

Published by Great Minds®.

Copyright © 2019 Great Minds®.

Printed in the U.S.A.

This book may be purchased from the publisher at eureka-math.org.

10   9   8   7   6   5   4   3   2

ISBN 978-1-64054-968-5

G6-M5-LPS-05.2019

Students, families, and educators:

Thank you for being part of the *Eureka Math®* community, where we celebrate the joy, wonder, and thrill of mathematics.

In *Eureka Math* classrooms, learning is activated through rich experiences and dialogue. That new knowledge is best retained when it is reinforced with intentional practice. The *Learn, Practice, Succeed* book puts in students' hands the problem sets and fluency exercises they need to express and consolidate their classroom learning and master grade-level mathematics. Once students learn and practice, they know they can succeed.

What is in the Learn, Practice, Succeed *book?*

Fluency Practice: Our printed fluency activities utilize the format we call a Sprint. Instead of rote recall, Sprints use patterns across a sequence of problems to engage students in reasoning and to reinforce number sense while building speed and accuracy. Sprints are inherently differentiated, with problems building from simple to complex. The tempo of the Sprint provides a low-stakes adrenaline boost that increases memory and automaticity.

Classwork: A carefully sequenced set of examples, exercises, and reflection questions support students' in-class experiences and dialogue. Having classwork preprinted makes efficient use of class time and provides a written record that students can refer to later.

Exit Tickets: Students show teachers what they know through their work on the daily Exit Ticket. This check for understanding provides teachers with valuable real-time evidence of the efficacy of that day's instruction, giving critical insight into where to focus next.

Homework Helpers and Problem Sets: The daily Problem Set gives students additional and varied practice and can be used as differentiated practice or homework. A set of worked examples, Homework Helpers, support students' work on the Problem Set by illustrating the modeling and reasoning the curriculum uses to build understanding of the concepts the lesson addresses.

Homework Helpers and Problem Sets from prior grades or modules can be leveraged to build foundational skills. When coupled with *Affirm®*, *Eureka Math's* digital assessment system, these Problem Sets enable educators to give targeted practice and to assess student progress. Alignment with the mathematical models and language used across *Eureka Math* ensures that students notice the connections and relevance to their daily instruction, whether they are working on foundational skills or getting extra practice on the current topic.

Where can I learn more about Eureka Math *resources?*

The Great Minds® team is committed to supporting students, families, and educators with an ever-growing library of resources, available at eureka-math.org. The website also offers inspiring stories of success in the *Eureka Math* community. Share your insights and accomplishments with fellow users by becoming a *Eureka Math* Champion.

Best wishes for a year filled with "aha" moments!

Jill Diniz

Jill Diniz
Chief Academic Officer, Mathematics
Great Minds

Contents

Module 5: Area, Surface Area, and Volume Problems

Number Correct: _____

Multiplication of Fractions I—Round 1

Directions: Determine the product of the fractions and simplify.

1.	$\dfrac{1}{2} \times \dfrac{3}{4}$		16.	$\dfrac{8}{9} \times \dfrac{3}{4}$	
2.	$\dfrac{5}{6} \times \dfrac{5}{7}$		17.	$\dfrac{3}{4} \times \dfrac{4}{7}$	
3.	$\dfrac{3}{4} \times \dfrac{7}{8}$		18.	$\dfrac{1}{4} \times \dfrac{8}{9}$	
4.	$\dfrac{4}{5} \times \dfrac{8}{9}$		19.	$\dfrac{3}{5} \times \dfrac{10}{11}$	
5.	$\dfrac{1}{4} \times \dfrac{3}{7}$		20.	$\dfrac{8}{13} \times \dfrac{7}{24}$	
6.	$\dfrac{5}{7} \times \dfrac{4}{9}$		21.	$2\dfrac{1}{2} \times 3\dfrac{3}{4}$	
7.	$\dfrac{3}{5} \times \dfrac{1}{8}$		22.	$1\dfrac{4}{5} \times 6\dfrac{1}{3}$	
8.	$\dfrac{2}{9} \times \dfrac{7}{9}$		23.	$8\dfrac{2}{7} \times 4\dfrac{5}{6}$	
9.	$\dfrac{1}{3} \times \dfrac{2}{5}$		24.	$5\dfrac{2}{5} \times 2\dfrac{1}{8}$	
10.	$\dfrac{3}{7} \times \dfrac{5}{8}$		25.	$4\dfrac{6}{7} \times 1\dfrac{1}{4}$	
11.	$\dfrac{2}{3} \times \dfrac{9}{10}$		26.	$2\dfrac{2}{3} \times 4\dfrac{2}{5}$	
12.	$\dfrac{3}{5} \times \dfrac{1}{6}$		27.	$6\dfrac{9}{10} \times 7\dfrac{1}{3}$	
13.	$\dfrac{2}{7} \times \dfrac{3}{4}$		28.	$1\dfrac{3}{8} \times 4\dfrac{2}{5}$	
14.	$\dfrac{5}{8} \times \dfrac{3}{10}$		29.	$3\dfrac{5}{6} \times 2\dfrac{4}{15}$	
15.	$\dfrac{4}{5} \times \dfrac{7}{8}$		30.	$4\dfrac{1}{3} \times 5$	

Number Correct: _____
Improvement: _____

Multiplication of Fractions I—Round 2

Directions: Determine the product of the fractions and simplify.

1.	$\dfrac{5}{6} \times \dfrac{1}{4}$	
2.	$\dfrac{2}{3} \times \dfrac{5}{7}$	
3.	$\dfrac{1}{3} \times \dfrac{2}{5}$	
4.	$\dfrac{5}{7} \times \dfrac{5}{8}$	
5.	$\dfrac{3}{8} \times \dfrac{7}{9}$	
6.	$\dfrac{3}{4} \times \dfrac{5}{6}$	
7.	$\dfrac{2}{7} \times \dfrac{3}{8}$	
8.	$\dfrac{1}{4} \times \dfrac{3}{4}$	
9.	$\dfrac{5}{8} \times \dfrac{3}{10}$	
10.	$\dfrac{6}{11} \times \dfrac{1}{2}$	
11.	$\dfrac{6}{7} \times \dfrac{5}{8}$	
12.	$\dfrac{1}{6} \times \dfrac{9}{10}$	
13.	$\dfrac{3}{4} \times \dfrac{8}{9}$	
14.	$\dfrac{5}{6} \times \dfrac{2}{3}$	
15.	$\dfrac{1}{4} \times \dfrac{8}{11}$	

16.	$\dfrac{3}{7} \times \dfrac{2}{9}$	
17.	$\dfrac{4}{5} \times \dfrac{10}{13}$	
18.	$\dfrac{2}{9} \times \dfrac{3}{8}$	
19.	$\dfrac{1}{8} \times \dfrac{4}{5}$	
20.	$\dfrac{3}{7} \times \dfrac{2}{15}$	
21.	$1\dfrac{1}{2} \times 4\dfrac{3}{4}$	
22.	$2\dfrac{5}{6} \times 3\dfrac{3}{8}$	
23.	$1\dfrac{7}{8} \times 5\dfrac{1}{5}$	
24.	$6\dfrac{2}{3} \times 2\dfrac{3}{8}$	
25.	$7\dfrac{1}{2} \times 3\dfrac{6}{7}$	
26.	$3 \times 4\dfrac{1}{3}$	
27.	$2\dfrac{3}{5} \times 5\dfrac{1}{6}$	
28.	$4\dfrac{2}{5} \times 7$	
29.	$1\dfrac{4}{7} \times 2\dfrac{1}{2}$	
30.	$3\dfrac{5}{6} \times \dfrac{3}{10}$	

Opening Exercise

Name each shape.

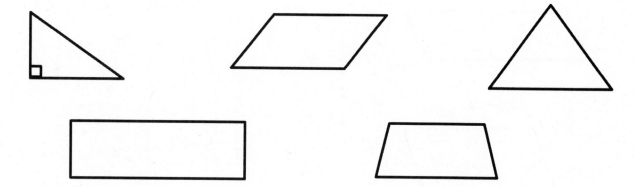

Exercises

1. Find the area of each parallelogram below. Note that the figures are not drawn to scale.

    a.

    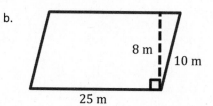
    4 cm
    5 cm
    6 cm

    b.

    8 m
    10 m
    25 m

    c.

    7 ft.
    11.5 ft.
    12 ft.

2. Draw and label the height of each parallelogram. Use the correct mathematical tool to measure (in inches) the base and height, and calculate the area of each parallelogram.

a.

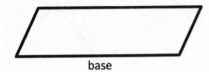

base

b.

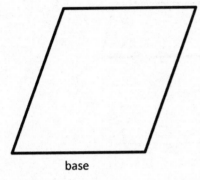

base

c.

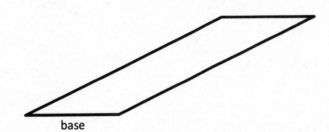

base

3. If the area of a parallelogram is $\frac{35}{42}$ cm^2 and the height is $\frac{1}{7}$ cm, write an equation that relates the height, base, and area of the parallelogram. Solve the equation.

EUREKA
MATH

Lesson Summary

The formula to calculate the area of a parallelogram is $A = bh$, where b represents the base and h represents the height of the parallelogram.

The height of a parallelogram is the line segment perpendicular to the base. The height is usually drawn from a vertex that is opposite the base.

Name _____   Date _____

Calculate the area of each parallelogram. Note that the figures are not drawn to scale.

1.

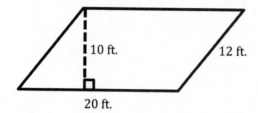

2.

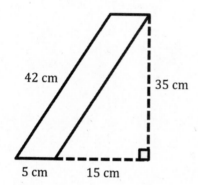

3.

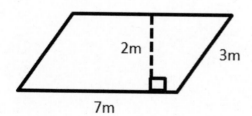

1. Draw and label the height of the parallelogram below.

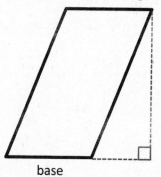

height

> I know the height represents a line segment that is perpendicular to the base and whose endpoint is on the opposite side of the parallelogram.

base

2. Calculate the area of this parallelogram. This **figure** is not drawn to scale.

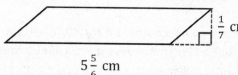

$\frac{1}{7}$ cm

$5\frac{5}{6}$ cm

> I know the base of the figure is $5\frac{5}{6}$ cm and the height of the figure is $\frac{1}{7}$ cm. I can substitute these values into the equation to calculate the area.

$$A = bh$$

$$= \left(5\frac{5}{6} \text{ cm}\right)\left(\frac{1}{7} \text{ cm}\right)$$

$$= \left(\frac{35}{6} \text{ cm}\right)\left(\frac{1}{7} \text{ cm}\right)$$

$$= \frac{35}{42} \text{ cm}^2$$

> In order to efficiently multiply these numbers, I rename $5\frac{5}{6}$ as a fraction greater than one. Since $\frac{6}{6}$ is equal to one, $\frac{30}{6}$ is equal to 5. When I add the remaining $\frac{5}{6}$, the resulting fraction is $\frac{35}{6}$.

3. Do the rectangle and parallelogram below have the same area? Explain why or why not. Note that the figures are not drawn to scale.

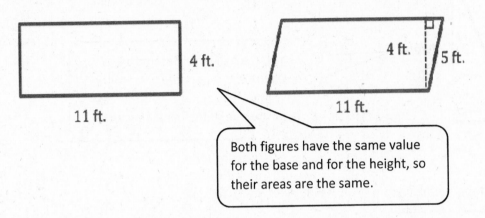

Both figures have the same value for the base and for the height, so their areas are the same.

Yes, the rectangle and parallelogram have the same area because I can decompose a right triangle on the right side of the parallelogram and move it over to the left side. This transforms the parallelogram into a rectangle. At this time, both rectangles would have the same dimensions; therefore, their areas are the same.

4. A parallelogram has an area of 57.76 square centimeters and a base of 15.2 centimeters. Write an equation that relates the area to the base and height, h. Solve the equation to determine the length of the height.

$$57.76 \, \text{cm}^2 = (15.2 \, \text{cm})h$$

$$57.76 \, \text{cm}^2 \div 15.2 \, \text{cm} = (15.2 \, \text{cm})h \div 15.2 \, \text{cm}$$

$$3.8 \, \text{cm} = h$$

The height of the parallelogram is 3.8 *cm.*

EUREKA MATH

Draw and label the height of each parallelogram.

1.

base

2.

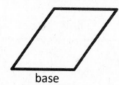

base

Calculate the area of each parallelogram. The figures are not drawn to scale.

3.

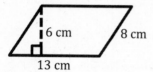

6 cm   8 cm

13 cm

4.

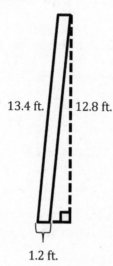

13.4 ft.   12.8 ft.

1.2 ft.

5.

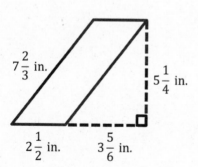

$7\frac{2}{3}$ in.

$5\frac{1}{4}$ in.

$2\frac{1}{2}$ in. $3\frac{5}{6}$ in.

6.

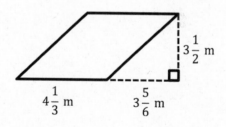

$3\frac{1}{2}$ m

$4\frac{1}{3}$ m $3\frac{5}{6}$ m

7. Brittany and Sid were both asked to draw the height of a parallelogram. Their answers are below.

Brittany

Sid

    Are both Brittany and Sid correct? If not, who is correct? Explain your answer.

8. Do the rectangle and parallelogram below have the same area? Explain why or why not.

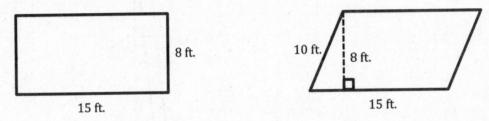

9. A parallelogram has an area of 20.3 cm^2 and a base of 2.5 cm. Write an equation that relates the area to the base and height, h. Solve the equation to determine the height of the parallelogram.

EUREKA
MATH®

Exploratory Challenge

a. Use the shapes labeled with an X to predict the formula needed to calculate the area of a right triangle. Explain your prediction.

Formula for the area of right triangles: _____

Area of the given triangle: _____

b. Use the shapes labeled with a Y to determine if the formula you discovered in part (a) is correct.

Does your area formula for triangle Y match the formula you got for triangle X?

If so, do you believe you have the correct formula needed to calculate the area of a right triangle? Why or why not?

If not, which formula do you think is correct? Why?

Area of the given triangle: _____

Exercises

Calculate the area of each triangle below. Each figure is not drawn to scale.

1.

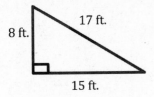

2.

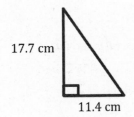

3.

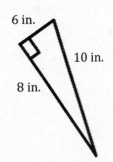

4.

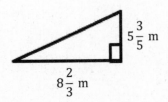

EUREKA
MATH

5.

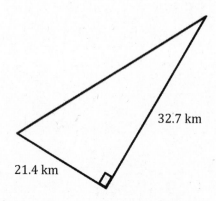

32.7 km

21.4 km

6. Mr. Jones told his students they each need half of a piece of paper. Calvin cut his piece of paper horizontally, and Matthew cut his piece of paper diagonally. Which student has the larger area on his half piece of paper? Explain.

Calvin's Paper

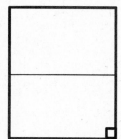

Matthew's Paper

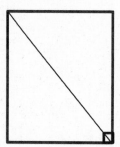

7. Ben requested that the rectangular stage be split into two equal sections for the upcoming school play. The only instruction he gave was that he needed the area of each section to be half of the original size. If Ben wants the stage to be split into two right triangles, did he provide enough information? Why or why not?

8. If the area of a right triangle is 6.22 sq. in. and its base is 3.11 in., write an equation that relates the area to the height, h, and the base. Solve the equation to determine the height.

Name _____   Date _____

1. Calculate the area of the right triangle. Each figure is not drawn to scale.

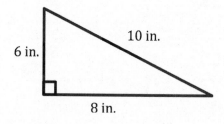

2. Dan and Joe are responsible for cutting the grass on the local high school soccer field. Joe cuts a diagonal line through the field, as shown in the diagram below, and says that each person is responsible for cutting the grass on one side of the line. Dan says that this is not fair because he will have to cut more grass than Joe. Is Dan correct? Why or why not?

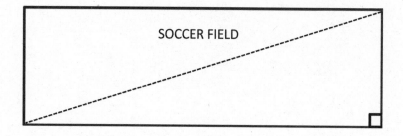

SOCCER FIELD

1. Calculate the area of the right triangle below. The figure is not drawn to scale.

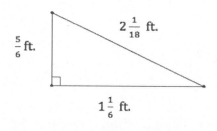

I can substitute the values for base and height into the formula for area of a triangle.

$$A = \frac{1}{2}bh = \frac{1}{2}\left(1\frac{1}{6}\text{ ft.}\right)\left(\frac{5}{6}\text{ ft.}\right) = \frac{1}{2}\left(\frac{7}{6}\text{ ft.}\right)\left(\frac{5}{6}\text{ ft.}\right) = \frac{35}{72}\text{ ft}^2$$

Before I multiply, I need to rename the mixed number as a fraction greater than one.

2. Elise has two rugs at her house. Both rugs have the same length and same width. Elise cut one rug horizontally across the middle, and she made a diagonal cut through the other rug.

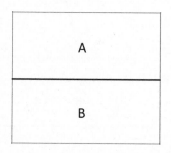

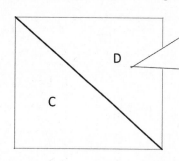

If I think about folding two identical pieces of paper to represent these two situations, each new section would represent exactly half of the original piece of paper.

After making the cuts, which rug (labeled A, B, C, or D) has the larger area? Explain.

After making the cuts, the new rugs are all the same size. The horizontal line goes through the center of the rectangle, making two equal parts. The diagonal line also splits the rectangle in two equal parts splits the rectangle in two equal parts because the area of a right triangle is exactly half the area of the rectangle.

3. Give the dimensions of a right triangle and a parallelogram with the same area. Explain how you know.

   A right triangle has a base of 10 cm and a height of 2 cm.

   $$A = \frac{1}{2}bh$$

   $$A = \frac{1}{2}(10\,\text{cm})(2\,\text{cm})$$

   $$A = 10\,\text{cm}^2$$

   The area of the triangle is 10 cm². In order to find the dimensions of a parallelogram that has the same area, I would need to find factors of 10 because I multiply the base and height of a parallelogram to find the area. Therefore, the parallelogram can have dimensions of 1 cm and 10 cm or 2 cm and 5 cm.

4. If the area of a right triangle is $\frac{7}{20}$ square feet and the height is $\frac{1}{5}$ feet, write an equation that relates the area to the base, b, and the height. Solve the equation to determine the base.

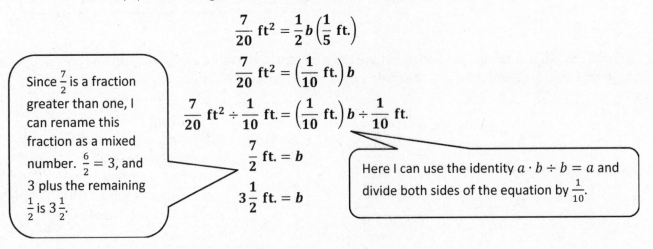

$$\frac{7}{20}\,\text{ft}^2 = \frac{1}{2}b\left(\frac{1}{5}\,\text{ft.}\right)$$

$$\frac{7}{20}\,\text{ft}^2 = \left(\frac{1}{10}\,\text{ft.}\right)b$$

$$\frac{7}{20}\,\text{ft}^2 \div \frac{1}{10}\,\text{ft.} = \left(\frac{1}{10}\,\text{ft.}\right)b \div \frac{1}{10}\,\text{ft.}$$

$$\frac{7}{2}\,\text{ft.} = b$$

$$3\frac{1}{2}\,\text{ft.} = b$$

Since $\frac{7}{2}$ is a fraction greater than one, I can rename this fraction as a mixed number. $\frac{6}{2} = 3$, and 3 plus the remaining $\frac{1}{2}$ is $3\frac{1}{2}$.

Here I can use the identity $a \cdot b \div b = a$ and divide both sides of the equation by $\frac{1}{10}$.

Therefore, the base of the right triangle is $3\frac{1}{2}$ ft.

EUREKA
MATH

Calculate the area of each right triangle below. Note that the figures are not drawn to scale.

1.

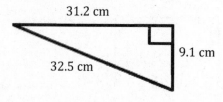

2.

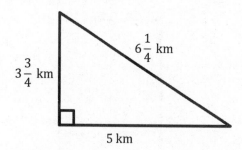

3.

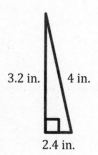

4.

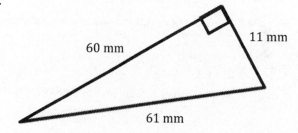

5.

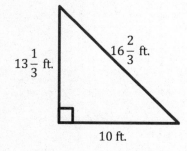

6. Elania has two congruent rugs at her house. She cut one vertically down the middle, and she cut diagonally through the other one.

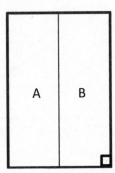 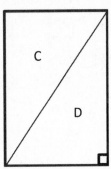

After making the cuts, which rug (labeled A, B, C, or D) has the larger area? Explain.

7. Give the dimensions of a right triangle and a parallelogram with the same area. Explain how you know.

8. If the area of a right triangle is $\frac{9}{16}$ sq. ft. and the height is $\frac{3}{4}$ ft., write an equation that relates the area to the base, b, and the height. Solve the equation to determine the base.

EUREKA
MATH®

Number Correct: _____

Multiplication of Decimals – Round 1

Directions: Evaluate each expression.

1.	5×1	
2.	5×0.1	
3.	5×0.01	
4.	5×0.001	
5.	2×4	
6.	0.2×4	
7.	0.02×4	
8.	0.002×4	
9.	3×3	
10.	3×0.3	
11.	3×0.03	
12.	0.1×0.8	
13.	0.01×0.8	
14.	0.1×0.08	
15.	0.01×0.08	
16.	0.3×0.2	
17.	0.03×0.2	
18.	0.02×0.3	
19.	0.02×0.03	
20.	0.2×0.2	
21.	0.02×0.2	
22.	0.2×0.02	

23.	5×3	
24.	5×0.3	
25.	0.5×3	
26.	0.3×0.5	
27.	9×2	
28.	0.2×9	
29.	0.9×2	
30.	0.2×0.9	
31.	4×0.4	
32.	0.4×0.4	
33.	0.04×0.4	
34.	0.8×0.6	
35.	0.8×0.06	
36.	0.006×0.8	
37.	0.006×0.08	
38.	0.7×0.9	
39.	0.07×0.9	
40.	0.9×0.007	
41.	0.09×0.007	
42.	1.2×0.7	
43.	1.2×0.07	
44.	0.007×0.12	

Number Correct: _____

Improvement: _____

Multiplication of Decimals – Round 2

Directions: Evaluate each expression.

1.	9×1	
2.	0.9×1	
3.	0.09×1	
4.	0.009×1	
5.	2×2	
6.	2×0.2	
7.	2×0.02	
8.	2×0.002	
9.	3×2	
10.	0.3×2	
11.	2×0.03	
12.	0.7×0.1	
13.	0.07×0.1	
14.	0.01×0.7	
15.	0.01×0.07	
16.	0.2×0.4	
17.	0.02×0.4	
18.	0.4×0.02	
19.	0.04×0.02	
20.	0.1×0.1	
21.	0.01×0.1	
22.	0.1×0.01	

23.	3×4	
24.	3×0.4	
25.	0.3×4	
26.	0.4×0.3	
27.	7×7	
28.	7×0.7	
29.	0.7×7	
30.	0.7×0.7	
31.	2×0.8	
32.	0.2×0.8	
33.	0.02×0.8	
34.	0.6×0.5	
35.	0.6×0.05	
36.	0.005×0.6	
37.	0.005×0.06	
38.	0.9×0.9	
39.	0.09×0.9	
40.	0.009×0.9	
41.	0.009×0.09	
42.	1.3×0.6	
43.	1.3×0.06	
44.	0.006×1.3	

Exercises

1. Work with a partner on the exercises below. Determine if the area formula $A = \frac{1}{2}bh$ is always correct. You may use a calculator, but be sure to record your work on your paper as well. Figures are not drawn to scale.

	Area of Two Right Triangles	Area of Entire Triangle
15 cm, 17.4 cm, 12 cm, 9 cm, 12.6 cm		
6.5 ft., 5.2 ft., 8 ft., 3.9 ft.		
$2\frac{5}{6}$ in., 2 in., $\frac{5}{6}$ in.		
34 m, 12 m, 32 m		

2. Can we use the formula $A = \frac{1}{2} \times \text{base} \times \text{height}$ to calculate the area of triangles that are not right triangles? Explain your thinking.

3. Examine the given triangle and expression.

    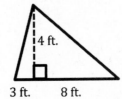

    $$\frac{1}{2}(11 \text{ ft.})(4 \text{ ft.})$$

    Explain what each part of the expression represents according to the triangle.

4. Joe found the area of a triangle by writing ⅃ $= \frac{1}{2}(11 \text{ in.})(4 \text{ in.})$, while Kaitlyn found the area by writing ⅃ $= \frac{1}{2}(3 \text{ in.})(4 \text{ in.}) + \frac{1}{2}(8 \text{ in.})(4 \text{ in.})$. Explain how each student approached the problem.

5. The triangle below has an area of 4.76 sq. in. If the base is 3.4 in., let h be the height in inches.

    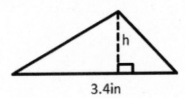

    a. Explain how the equation $4.76 \text{ in}^2 = \frac{1}{2}(3.4 \text{ in.})h$ represents the situation.

    b. Solve the equation.

EUREKA
MATH

Name _____   Date _____

Calculate the area of each triangle using two different methods. Figures are not drawn to scale.

1.

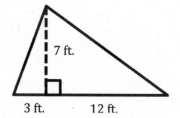

7 ft.

3 ft.   12 ft.

2.

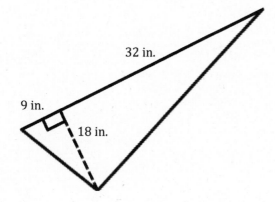

32 in.

9 in.

18 in.

Calculate the area of each shape below. Figures are not drawn to scale.

1.

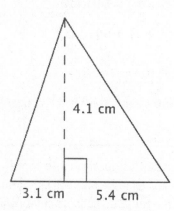

$$A = \frac{1}{2}(3.1\,\text{cm})(4.1\,\text{cm}) = 6.355\,\text{cm}^2$$

$$A = \frac{1}{2}(5.4\,\text{cm})(4.1\,\text{cm}) = 11.07\,\text{cm}^2$$

$$A = 6.355\,\text{cm}^2 + 11.07\,\text{cm}^2 = 17.425\,\text{cm}^2$$

OR

$$A = \frac{1}{2}(8.5\,\text{cm})(4.1\,\text{cm}) = 17.425\,\text{cm}^2$$

I can decompose the large triangle into two smaller triangles and find the area of each smaller triangle using the formula $A = \frac{1}{2}bh$ and then add the two areas. The sum of both areas is the total area of the triangle. Or, I can find the length of the base by adding 3.1 cm and 5.4 cm and then use the formula to determine the total area.

I can decompose the trapezoid into two right triangles and a square. I can calculate the area of each of these smaller shapes and then find the sum of these areas, which is the total area of the figure.

2.

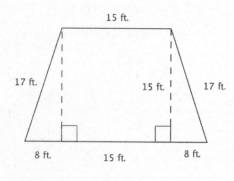

$$A = \frac{1}{2}(8\,\text{ft.})(15\,\text{ft.}) = 60\,\text{ft}^2$$

$$A = (15\,\text{ft.})(15\,\text{ft.}) = 225\,\text{ft}^2$$

$$A = \frac{1}{2}(8\,\text{ft.})(15\,\text{ft.}) = 60\,\text{ft}^2$$

$$A = 60\,\text{ft}^2 + 225\,\text{ft}^2 + 60\,\text{ft}^2 = 345\,\text{ft}^2$$

3.

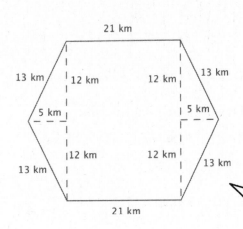

$$A = \frac{1}{2}(24\,\text{km})(5\,\text{km}) = 60\,\text{km}^2$$

$$A = (21\,\text{km})(24\,\text{km}) = 504\,\text{km}^2$$

$$A = \frac{1}{2}(24\,\text{km})(5\,\text{km}) = 60\,\text{km}^2$$

$$A = 60\,\text{km}^2 + 504\,\text{km}^2 + 60\,\text{km}^2 = 624\,\text{km}^2$$

I can decompose this hexagon into three parts. There is a triangle on each side and a rectangle in the middle. I can calculate the area of the entire figure by finding the sum of the areas of the smaller shapes.

4. Jasmine is building an enclosure for her rabbits. The bottom of the enclosure is in the shape of a triangle, with a base of 72 inches and an altitude of 36 inches. How much space will her rabbits have?

$$A = \frac{1}{2}bh = \frac{1}{2}(72\,\text{in.})(36\,\text{in.}) = 1{,}296\,\text{in}^2$$

Jasmine's rabbits will have 1,296 in² in which to play.

By calculating the area of the space, I can determine how much room there is inside the enclosure, which is where the rabbits will play.

5. Examine the triangle to the right.

   a. Write an expression to show how you would calculate the area.

$$\frac{1}{2}(9\,\text{in.})(4\,\text{in.}) + \frac{1}{2}(3\,\text{in.})(4\,\text{in.}) \ \text{or} \ \frac{1}{2}(12\,\text{in.})(4\,\text{in.})$$

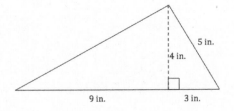

   b. Identify each part of your expression as it relates to the triangle.

   In the first expression, I decomposed the larger triangle into two smaller triangles. The base is composed of 9 in. and 3 in., and the height, or the altitude, is 4 in. The sum of the areas of these two triangles is the total area of the triangle.

   In the second expression, the base of the larger triangle is 12 in. because 9 in. + 3 in. = 12 in., and the height, or the altitude, is 4 in.

EUREKA MATH

6. A room has a triangular floor with an area of $16\frac{1}{4}$ sq. m. If the altitude of the triangle is $3\frac{1}{2}$ m, write an equation to determine the length of the base, b, in meters. Then solve the equation.

> To solve for b, I can use the identity $a \cdot b \div b = a$.

> I can substitute the given values for area and height into the formula and rename the mixed numbers as fractions greater than one to efficiently multiply.

$$16\frac{1}{4}\ \text{m}^2 = \frac{1}{2}b\left(3\frac{1}{2}\ \text{m}\right)$$

$$\frac{65}{4}\ \text{m}^2 = \left(\frac{7}{4}\ \text{m}\right)b$$

$$\frac{65}{4}\ \text{m}^2 \div \frac{7}{4}\ \text{m} = \left(\frac{7}{4}\ \text{m}\right)b \div \frac{7}{4}\ \text{m}$$

$$\frac{65}{7}\ \text{m} = b$$

$$9\frac{2}{7}\ \text{m} = b$$

> After dividing the fractions greater than one, I can rename the fraction greater than one, $\frac{65}{7}$, as a mixed number.

Calculate the area of each shape below. Figures are not drawn to scale.

1.

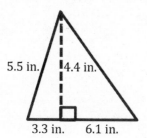

5.5 in. 4.4 in.

3.3 in. 6.1 in.

2.

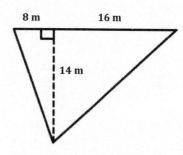

8 m   16 m

14 m

3.

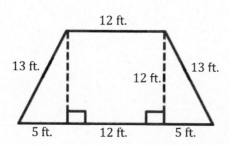

12 ft.

13 ft.   13 ft.

12 ft.

5 ft. 12 ft. 5 ft.

4.

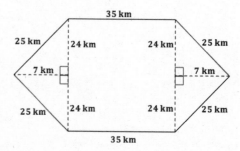

35 km

25 km   24 km   24 km   25 km

7 km   7 km

25 km   24 km   24 km   25 km

35 km

5. Immanuel is building a fence to make an enclosed play area for his dog. The enclosed area will be in the shape of a triangle with a base of 48 m. and an altitude of 32 m. How much space does the dog have to play?

6. Chauncey is building a storage bench for his son's playroom. The storage bench will fit into the corner and against two walls to form a triangle. Chauncey wants to buy a triangular shaped cover for the bench.

If the storage bench is $2\frac{1}{2}$ ft. along one wall and $4\frac{1}{4}$ ft. along the other wall, how big will the cover have to be to cover the entire bench?

Note: Figure is not to scale.

7. Examine the triangle to the right.

a. Write an expression to show how you would calculate the area.

b. Identify each part of your expression as it relates to the triangle.

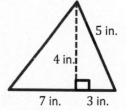

5 in.

4 in.

7 in. 3 in.

8. The floor of a triangular room has an area of $32\frac{1}{2}$ sq. m. If the triangle's altitude is $7\frac{1}{2}$ m, write an equation to determine the length of the base, b, in meters. Then solve the equation.

Opening Exercise

Draw and label the altitude of each triangle below.

a.

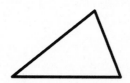

b.

c.

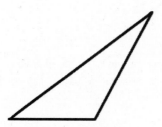

Exploratory Challenge/Exercises 1–5

1. Use rectangle X and the triangle with the altitude inside (triangle X) to show that the area formula for the triangle is $A = \frac{1}{2} \times \text{base} \times \text{height}$.

    a. Step One: Find the area of rectangle X.

    b. Step Two: What is half the area of rectangle X?

c. Step Three: Prove, by decomposing triangle X, that it is the same as half of rectangle X. Please glue your decomposed triangle onto a separate sheet of paper. Glue it into rectangle X. What conclusions can you make about the triangle's area compared to the rectangle's area?

2. Use rectangle Y and the triangle with a side that is the altitude (triangle Y) to show the area formula for the triangle is $A = \frac{1}{2} \times$ base $\times$ height.

a. Step One: Find the area of rectangle Y.

b. Step Two: What is half the area of rectangle Y?

c. Step Three: Prove, by decomposing triangle Y, that it is the same as half of rectangle Y. Please glue your decomposed triangle onto a separate sheet of paper. Glue it into rectangle Y. What conclusions can you make about the triangle's area compared to the rectangle's area?

3. Use rectangle Z and the triangle with the altitude outside (triangle Z) to show the area formula for the triangle is $A = \frac{1}{2} \times$ base $\times$ height.

a. Step One: Find the area of rectangle Z.

b. Step Two: What is half the area of rectangle Z?

c. Step Three: Prove, by decomposing triangle Z, that it is the same as half of rectangle Z. Please glue your decomposed triangle onto a separate sheet of paper. Glue it into rectangle Z. What conclusions can you make about the triangle's area compared to the rectangle's area?

Lesson 4:   The Area of All Triangles Using Height and Base

EUREKA MATH®

4.   When finding the area of a triangle, does it matter where the altitude is located?

5.   How can you determine which part of the triangle is the base and which is the height?

Exercises 6–8

Calculate the area of each triangle. Figures are not drawn to scale.

6.

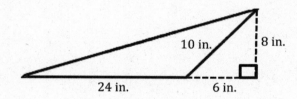

7.

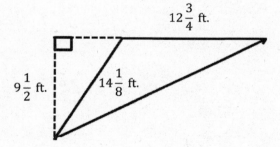

8.   Draw three triangles (acute, right, and obtuse) that have the same area. Explain how you know they have the same area.

Name _____   Date _____

Find the area of each triangle. Figures are not drawn to scale.

1.

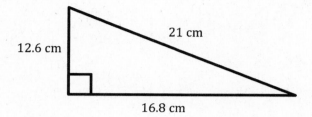

21 cm

12.6 cm

16.8 cm

2.

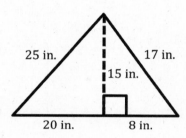

25 in.   17 in.

15 in.

20 in.   8 in.

3.

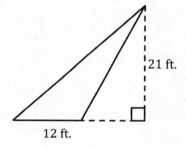

21 ft.

12 ft.

1. Calculate the area of this triangle. This figure is not drawn to scale.

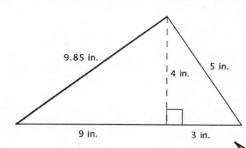

$$A = \frac{1}{2}(12\,\text{in.})(4\,\text{in.}) = 24\,\text{in}^2$$

To determine the area of the triangle, I need to know the base and height. The base of the triangle is the sum of 9 in. and 3 in., which is 12 in. The height is 4 in.

2. Calculate the area of this triangle. This figure is not drawn to scale.

The height of the obtuse triangle is found outside of the triangle.

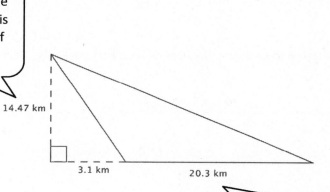

$$A = \frac{1}{2}(20.3\,\text{km})(14.47\,\text{km}) = 146.8705\,\text{km}^2$$

When I determine the length of the base, I remember not to use the length between the base and the altitude.

> I need to decompose the given figure in order to calculate the total area.

3. Calculate the area of the figure below. This figure is not drawn to scale.

> I can determine the total length of the base of the triangle in the upper region of this figure by adding 6 cm + 9 cm + 6 cm.

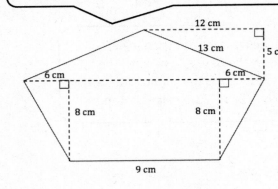

$$A = \frac{1}{2}(6 \text{ cm})(8 \text{ cm}) = 24 \text{ cm}^2$$

$$A = \frac{1}{2}(21 \text{ cm})(5 \text{ cm}) = 52.5 \text{ cm}^2$$

$$A = (9 \text{ cm})(8 \text{ cm}) = 72 \text{ cm}^2$$

$$A = 24 \text{ cm}^2 + 24 \text{ cm}^2 + 52.5 \text{ cm}^2 + 72 \text{ cm}^2$$

$$= 172.5 \text{ cm}^2$$

4. Jason and Joelle are both trying to calculate the area of an obtuse triangle. Examine their calculations below.

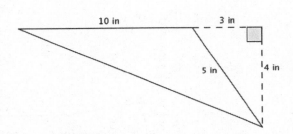

Jason's Work	Joelle's Work
$A = \frac{1}{2} \times 10 \text{ in.} \times 4 \text{ in.}$ $A = 20 \text{ in}^2$	$A = \frac{1}{2} \times 3 \text{ in.} \times 4 \text{ in.}$ $A = 6 \text{ in}^2$

Which student calculated the area correctly? Explain why the other student is not correct.

Jason calculated the area correctly. Although Joelle did use the altitude of the triangle, she used the length between the altitude and the base rather than the length of the actual base.

Lesson 4: The Area of All Triangles Using Height and Base

EUREKA MATH

5. David calculated the area of the triangle below. His work is shown.

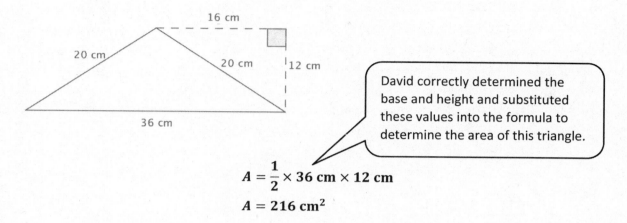

David correctly determined the base and height and substituted these values into the formula to determine the area of this triangle.

$$A = \frac{1}{2} \times 36 \text{ cm} \times 12 \text{ cm}$$

$$A = 216 \text{ cm}^2$$

Although David was told his work is correct, he had a hard time explaining why it is correct. Help David explain why his calculations are correct.

The formula for the area of a triangle is $A = \frac{1}{2}bh$. David followed this formula because 12 cm is the height of the triangle, and 36 cm is the base of the triangle.

6. The larger triangle below has a base of 15.47 m; the gray triangle has an area of 56.172 m².

   a. Determine the area of the larger triangle if it has a height of 11.3 m.

$$A = \frac{1}{2}(15.47 \text{ m})(11.3 \text{ m})$$

$$A = 87.4055 \text{ m}^2$$

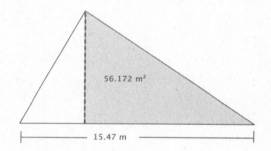

   b. Let A be the area of the unshaded (white) triangle in square meters. Write and solve an equation to determine the value of A using the areas of the larger triangle and the gray triangle.

$$56.172 \text{ m}^2 + A = 87.4055 \text{ m}^2$$

$$56.172 \text{ m}^2 + A - 56.172 \text{ m}^2 = 87.4055 \text{ m}^2 - 56.172 \text{ m}^2$$

$$A = 31.2335 \text{ m}^2$$

The sum of the area of the white triangle and the area of the gray triangle is equal to the area of the larger triangle, 87.4055 m². I know the area of the gray triangle is 56.172 m².

Calculate the area of each figure below. Figures are not drawn to scale.

1.

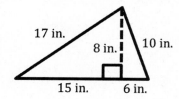

2.

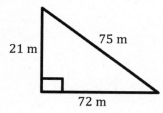

3.

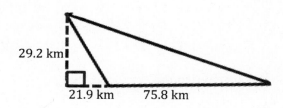

4.

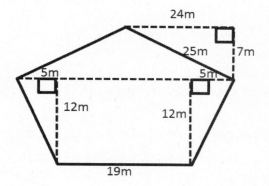

5. The Andersons are going on a long sailing trip during the summer. However, one of the sails on their sailboat ripped, and they have to replace it. The sail is pictured below.

If the sailboat sails are on sale for $2 per square foot, how much will the new sail cost?

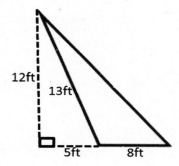

6. Darnell and Donovan are both trying to calculate the area of an obtuse triangle. Examine their calculations below.

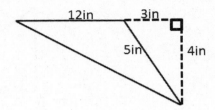

Darnell's Work	Donovan's Work
$\wr = \dfrac{1}{2} \times 3 \text{ in.} \times 4 \text{ in.}$ $\wr = 6 \text{ in}^2$	$\wr = \dfrac{1}{2} \times 12 \text{ in.} \times 4 \text{ in.}$ $\wr = 24 \text{ in}^2$

Which student calculated the area correctly? Explain why the other student is not correct.

7. Russell calculated the area of the triangle below. His work is shown.

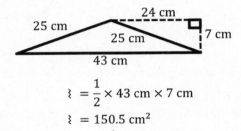

$$\wr = \frac{1}{2} \times 43 \text{ cm} \times 7 \text{ cm}$$

$$\wr = 150.5 \text{ cm}^2$$

Although Russell was told his work is correct, he had a hard time explaining why it is correct. Help Russell explain why his calculations are correct.

8. The larger triangle below has a base of 10.14 m; the gray triangle has an area of 40.325 m².

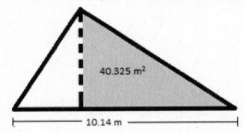

a. Determine the area of the larger triangle if it has a height of 12.2 m.

b. Let $\wr$ be the area of the unshaded (white) triangle in square meters. Write and solve an equation to determine the value of $\wr$, using the areas of the larger triangle and the gray triangle.

Opening Exercise

Here is an aerial view of a woodlot.

If $AB = 10$ units, $FE = 8$ units, $AF = 6$ units, and $DE = 7$ units, find the lengths of the other two sides.

$DC =$

$BC =$

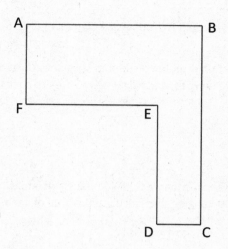

If $DC = 10$ units, $FE = 30$ units, $AF = 28$ units, and $BC = 54$ units, find the lengths of the other two sides.

$AB =$

$DE =$

Discussion

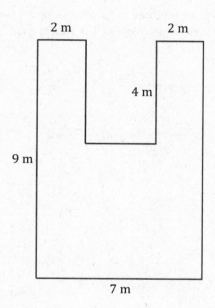

Example 1: Decomposing Polygons into Rectangles

The Intermediate School is producing a play that needs a special stage built. A diagram of the stage is shown below (not to scale).

a. On the first diagram, divide the stage into three rectangles using two horizontal lines. Find the dimensions of these rectangles, and calculate the area of each. Then, find the total area of the stage.

b. On the second diagram, divide the stage into three rectangles using two vertical lines. Find the dimensions of these rectangles, and calculate the area of each. Then, find the total area of the stage.

c. On the third diagram, divide the stage into three rectangles using one horizontal line and one vertical line. Find the dimensions of these rectangles, and calculate the area of each. Then, find the total area of the stage.

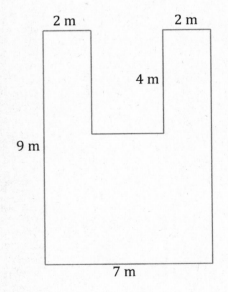
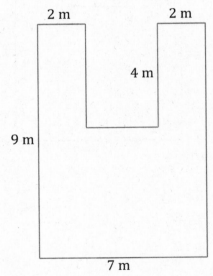
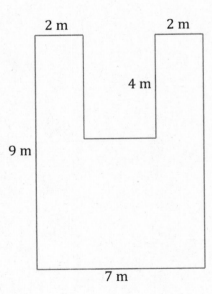

EUREKA
MATH

d. Think of this as a large rectangle with a piece removed.

i. What are the dimensions of the large rectangle and the small rectangle?

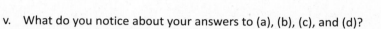

ii. What are the areas of the two rectangles?

iii. What operation is needed to find the area of the original figure?

iv. What is the difference in area between the two rectangles?

v. What do you notice about your answers to (a), (b), (c), and (d)?

vi. Why do you think this is true?

Example 2: Decomposing Polygons into Rectangles and Triangles

Parallelogram *ABCD* is part of a large solar power collector. The base measures 6 m and the height is 4 m.

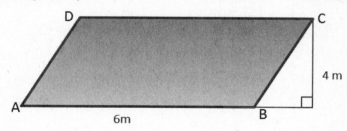

a. Draw a diagonal from *A* to *C*. Find the area of both triangles *ABC* and *ACD*.

b. Draw in the other diagonal, from *B* to *D*. Find the area of both triangles *ABD* and *BCD*.

Example 3: Decomposing Trapezoids

The trapezoid below is a scale drawing of a garden plot.

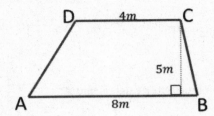

Find the area of both triangles ABC and ACD. Then find the area of the trapezoid.

Find the area of both triangles ABD and BCD. Then find the area of the trapezoid.

How else could we find this area?

EUREKA
MATH

Name _____   Date _____

1. Find the missing dimensions of the figure below, and then find the area. The figure is not drawn to scale.

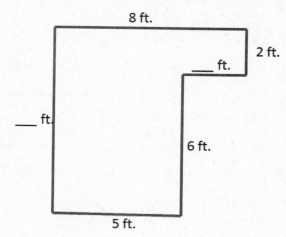

8 ft.

2 ft.

___ ft.

___ ft.

6 ft.

5 ft.

2. Find the area of the parallelogram below by decomposing into two triangles. The figure is not drawn to scale.

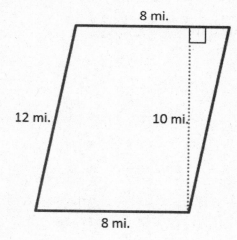

8 mi.

12 mi.   10 mi.

8 mi.

1. If $AB = 30$ units, $FE = 21$ units, $AF = 18$ units, and $DE = 21$ units, find the length of both other sides. Then, find the area of the irregular polygon. All measurements are in units.

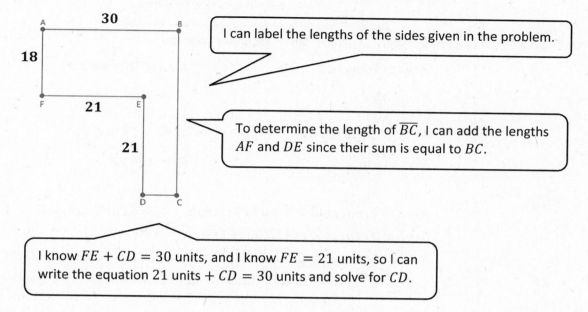

I can label the lengths of the sides given in the problem.

To determine the length of $\overline{BC}$, I can add the lengths AF and DE since their sum is equal to BC.

I know $FE + CD = 30$ units, and I know $FE = 21$ units, so I can write the equation 21 units $+ CD = 30$ units and solve for CD.

$CD = 30$ units $- 21$ units $= 9$ units

$BC = 18$ units $+ 21$ units $= 39$ units

To calculate the area of the figure, decompose the figure into two rectangles. The dimensions of one rectangle are 30 units × 18 units, and the dimensions of the other rectangle are 9 units × 21 units.

$A = 30$ units $\times 18$ units $= 540$ units^2

$A = 9$ units $\times 21$ units $= 189$ units^2

Total Area $= 540$ units$^2 + 189$ units$^2 = 729$ units^2

I can decompose this trapezoid into two triangles.

2. Determine the area of the trapezoid below. The trapezoid is not drawn to scale.

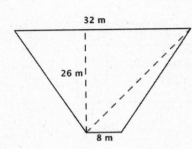

Area of Triangle 1

$A = \frac{1}{2}bh$

$A = \frac{1}{2} \times 32\,\text{m} \times 26\,\text{m}$

$A = 416\,\text{m}^2$

Area of Triangle 2

$A = \frac{1}{2}bh$

$A = \frac{1}{2} \times 8\,\text{m} \times 26\,\text{m}$

$A = 104\,\text{m}^2$

Area of Trapezoid = Area of Triangle 1 + Area of Triangle 2

Area of Trapezoid = 416 m² + 104 m² = 520 m²

I construct a rectangle around this isosceles trapezoid. To determine the base of each triangle, I subtract 4 m from 21 m and then divide the difference by 2 since the triangles have the same length for the base.

3. Determine the area of the isosceles trapezoid below. The image is not drawn to scale.

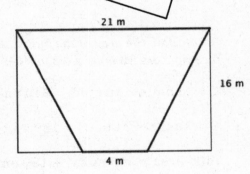

Area of Rectangle

$A = bh$

$A = 21\,\text{m} \times 16\,\text{m}$

$A = 336\,\text{m}^2$

Area of Triangles 1 and 2

$A = \frac{1}{2}bh$

$A = \frac{1}{2} \times 8.5\,\text{m} \times 16\,\text{m}$

$A = 68\,\text{m}^2$

Area of Trapezoid = Area of Rectangle − Area of Triangle 1 − Area of Triangle 2

Area of Trapezoid = 336 m² − 68 m² − 68 m² = 200 m²

4. Here is a sketch of a wall that needs to be painted.

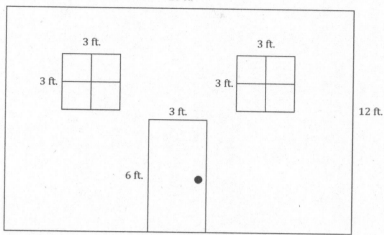

a. The windows and door will not be painted. Calculate the area of the wall that will be painted.

Area of the entire wall: $18 \text{ ft.} \times 12 \text{ ft.} = 216 \text{ ft}^2$

Area of the two windows: $2(3 \text{ ft.} \times 3 \text{ ft.}) = 18 \text{ ft}^2$

Area of the door: $6 \text{ ft.} \times 3 \text{ ft.} = 18 \text{ ft}^2$

Area that needs to be painted: $216 \text{ ft}^2 - 18 \text{ ft}^2 - 18 \text{ ft}^2 = 180 \text{ ft}^2$

The area of the wall that will be painted is 180 ft^2.

> To find the area of the wall that will be painted, I can determine the area of each part that will not be painted (the windows and the door) and subtract the areas of these parts from the area of the entire wall.

b. If a quart of Cover-All Paint covers 45 ft^2, how many quarts must be purchased for the painting job?

$$180 \div 45 = 4$$

Therefore, 4 quarts of paint must be purchased.

> Since 180 ft^2 is the total area that needs to be painted, I can divide this amount by the area covered by one quart.

EUREKA MATH   **Lesson 5:**   The Area of Polygons Through Composition   **59**
                                      and Decomposition

1. If $AB = 20$ units, $FE = 12$ units, $AF = 9$ units, and $DE = 12$ units, find the length of the other two sides. Then, find the area of the irregular polygon.

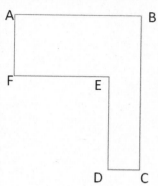

2. If $DC = 1.9$ cm, $FE = 5.6$ cm, $AF = 4.8$ cm, and $BC = 10.9$ cm, find the length of the other two sides. Then, find the area of the irregular polygon.

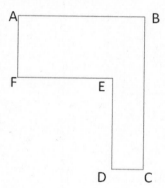

3. Determine the area of the trapezoid below. The trapezoid is not drawn to scale.

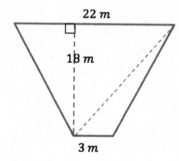

22 m

18 m

3 m

4. Determine the area of the shaded isosceles trapezoid below. The image is not drawn to scale.

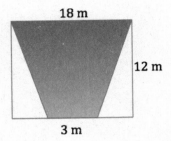

5. Here is a sketch of a wall that needs to be painted:

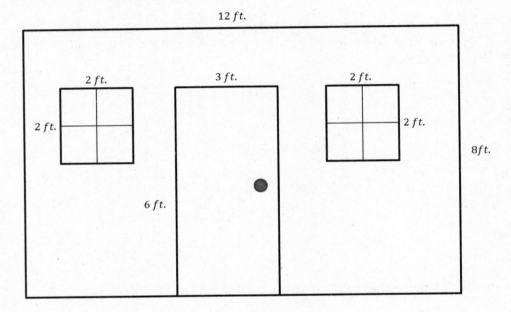

a. The windows and door will not be painted. Calculate the area of the wall that will be painted.

b. If a quart of Extra-Thick Gooey Sparkle paint covers 30 ft², how many quarts must be purchased for the painting job?

Lesson 5:   The Area of Polygons Through Composition
              and Decomposition

EUREKA
MATH®

Exploratory Challenge 1: Classroom Wall Paint

The custodians are considering painting our classroom next summer. In order to know how much paint they must buy, the custodians need to know the total surface area of the walls. Why do you think they need to know this, and how can we find the information?

Make a prediction of how many square feet of painted surface there are on one wall in the room. If the floor has square tiles, these can be used as a guide.

Estimate the dimensions and the area. Predict the area before you measure.

My prediction: _____ ft^2.

   a. Measure and sketch one classroom wall. Include measurements of windows, doors, or anything else that would not be painted.

   Sketch:

Object or Item to Be Measured	Measurement Units	Precision (measure to the nearest)	Length	Width	Expression that Shows the Area	Area
door	feet	half foot	$6\frac{1}{2}$ ft.	$3\frac{1}{2}$ ft.	$6\frac{1}{2}$ ft. $\times$ $3\frac{1}{2}$ ft.	$22\frac{3}{4}$ ft^2

b. Work with your partners and your sketch of the wall to determine the area that needs paint. Show your sketch and calculations below; clearly mark your measurements and area calculations.

c. A gallon of paint covers about 350 ft^2. Write an expression that shows the total area of the wall. Evaluate it to find how much paint is needed to paint the wall.

d. How many gallons of paint would need to be purchased to paint the wall?

Lesson 6:   Area in the Real World

EUREKA MATH

Exploratory Challenge 2

Object or Item to Be Measured	Measurement Units	Precision (measure to the nearest)	Length	Width	Area
door	feet	half foot	$6\frac{1}{2}$ ft.	$3\frac{1}{2}$ ft.	$22\frac{3}{4}$ ft^2

Name _____ Date _____

Find the area of the deck around this pool. The deck is the white area in the diagram.

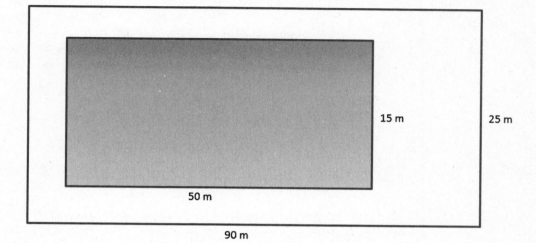

Lesson 6:   Area in the Real World

© 2019 Great Minds®. eureka-math.org

1. Below is a drawing of a wall that will be covered in either wallpaper or paint. The wall is 12 ft. high and 23 ft. long. The window, brick, and door will not be painted. The brick measures 7 ft. × 36 in., the door is 3 ft. wide and 6 ft. high, and the window is 8 ft. × 9 ft.

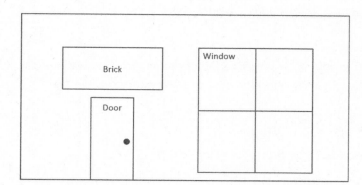

> I can label the dimensions of the wall, window, brick, and door. I notice the width of the brick is written in inches (36 in.), so I can convert that measurement to 3 ft. so all the units are feet.

a. How many square feet of wallpaper are needed to cover the wall?

Area of entire wall: **23 ft. × 12 ft. = 276 ft²**

Area of the brick: **7 ft. × 3 ft. = 21 ft²**

Area of the door: **3 ft. × 6 ft. = 18 ft²**

Area of the window: **8 ft. × 9 ft. = 72 ft²**

Area that will be painted: **276 ft² − (18 ft² + 21 ft² + 72 ft²) = 165 ft²**

> I can find the area of each part of the wall that is not going to be covered and add the areas together. Then I can subtract that total from the total area of the wall.

b. The wallpaper is sold in rolls that are 18 in. wide and 30 ft. long. Rolls of solid color wallpaper will be used, so patterns do not have to match up.

i. What is the area of one roll of wallpaper?

Area of one roll of wallpaper: **1.5 ft. × 30 ft. = 45 ft²**

> I have to pay close attention to the units.

ii. How many rolls would be needed to cover the wall?

165 ft² ÷ 45 ft² ≈ 3.7

It is necessary to buy 4 rolls of wallpaper.

> I have to round up because I can't purchase part of a roll of wallpaper.

c. This week, the rolls of wallpaper are on sale for $9.99 per roll. Find the cost of covering the wall with wallpaper.

$$\$9.99 \times 4 = \$39.96$$

We need four rolls of wallpaper to cover the wall, which will cost $39.96.

d. A gallon of special textured paint covers 200 ft^2 and is on sale for $19.99 per gallon. The wall needs to be painted twice (the wall needs two coats of paint). Find the cost of using paint to cover the wall.

$$165 \text{ ft}^2 \times 2 = 330 \text{ ft}^2$$

If the wall needs to be painted twice, we need to paint a total area of 330 ft^2.

$$330 \text{ ft}^2 \div 200 \text{ ft}^2 = 1.65$$

Two gallons of paint need to be purchased in order to paint the wall with two coats of paint.

$$\$19.99 \times 2 = \$39.98$$

The cost for two coats of paint is $39.98.

2. A classroom has a length of 27 ft. and a width of 32 ft. The flooring is to be replaced by tiles. If each tile has a length of 18 in. and a width of 36 in., how many tiles are needed to cover the classroom floor?

Area of the classroom: 27 ft. × 32 ft. = 864 ft^2

Area of each tile: 1.5 ft. × 3 ft. = 4.5 ft^2

To find the area of one tile, I can convert the dimensions from inches to feet and then multiply.

$$\frac{\textbf{Area of the classroom}}{\textbf{Area of each tile}} = \frac{864 \text{ ft}^2}{4.5 \text{ ft}^2} = 192$$

192 tiles are needed to cover the classroom floor.

EUREKA
MATH®

3. Challenge: Assume that the tiles from Problem 2 are unavailable. Another design is available, but the tiles are square, 18 in. on a side. If these are to be installed, how many must be ordered?

Solutions will vary. An even number of tiles fit on the 27 foot length of the room (18 tiles), but the width requires $21\frac{1}{3}$ tiles.

> I know the length of the room is 27 ft., so I can divide 27 by 1.5 since the length of one 18 in. tile is 1.5 ft. Since I know the width of the room is 32 ft., I can divide 32 by 1.5.

This accounts for an 18 tile by 21 tile array. $18 \times 21 = 378$, so 378 tiles must be ordered.

The remaining area is 27 ft. × 0.5 ft. (18 tiles × $\frac{1}{3}$ tile)

> The remaining area is the part in the upper portion of the image to the right. The length is 27 ft. (18 tiles), and the width is $\frac{1}{3}$ tile. Since the tile is 1.5 ft. long, $\frac{1}{3}$ of 1.5 ft. is 0.5 ft.

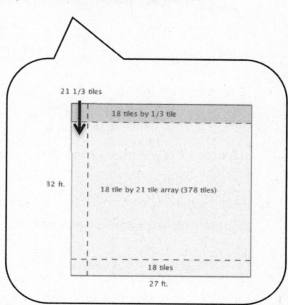

Since 18 of the $\frac{1}{3}$ tiles are needed, 6 additional tiles must be cut because $\frac{18}{3} = 6$.

Using the same logic as above, some students may correctly say they will need 384 tiles.

> I can add the number of tiles in the 18 tile by 21 tile array, 378 tiles, with the number of tiles needed for the remaining part, 6 tiles. 378 tiles + 6 tiles = 384 tiles.

4. Henry's deck has an open area where he would like to plant a vegetable garden.

   a. Find the missing portion of the deck. Write the expression and evaluate it.

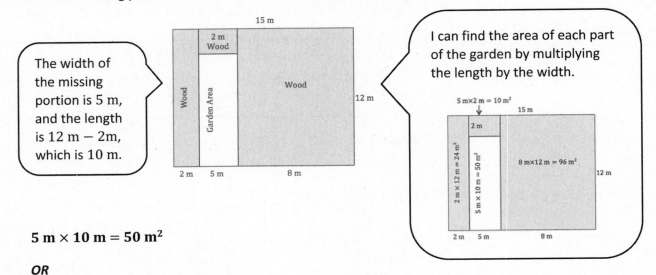

> The width of the missing portion is 5 m, and the length is 12 m − 2m, which is 10 m.

> I can find the area of each part of the garden by multiplying the length by the width.

$$5 \text{ m} \times 10 \text{ m} = 50 \text{ m}^2$$

OR

$$15 \text{ m} \times 12 \text{ m} - 2 \text{ m} \times 12 \text{ m} - 5 \text{ m} \times 2 \text{ m} - 8 \text{ m} \times 12 \text{ m} = 50 \text{ m}^2$$

   b. Find the missing portion of the deck using a different method. Write the expression and evaluate it. *Choose whichever method was not used in part (a).*

   c. Write your two equivalent expressions.

$$5 \text{ m} \times 10 \text{ m}$$
$$15 \text{ m} \times 12 \text{ m} - 2 \text{ m} \times 12 \text{ m} - 5 \text{ m} \times 2 \text{ m} - 8 \text{ m} \times 12 \text{ m}$$

   d. Explain how each demonstrates a different understanding of the diagram.
One expression shows the dimensions of the garden area (interior rectangle, $5 \text{ m} \times 10 \text{ m}$), and one shows finding the total area minus each of the three wooden areas.

EUREKA MATH

1. Below is a drawing of a wall that is to be covered with either wallpaper or paint. The wall is 8 ft. high and 16 ft. wide. The window, mirror, and fireplace are not to be painted or papered. The window measures 18 in. wide and 14 ft. high. The fireplace is 5 ft. wide and 3 ft. high, while the mirror above the fireplace is 4 ft. wide and 2 ft. high. (Note: this drawing is not to scale.)

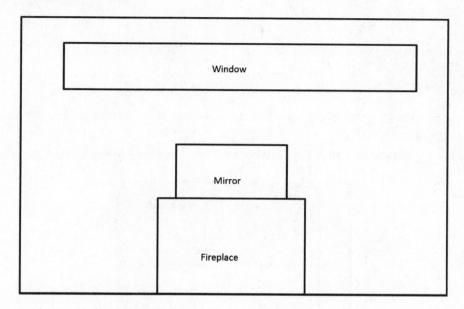

a. How many square feet of wallpaper are needed to cover the wall?

b. The wallpaper is sold in rolls that are 18 in. wide and 33 ft. long. Rolls of solid color wallpaper will be used, so patterns do not have to match up.

    i.   What is the area of one roll of wallpaper?

    ii. How many rolls would be needed to cover the wall?

c. This week, the rolls of wallpaper are on sale for $11.99/roll. Find the cost of covering the wall with wallpaper.

d. A gallon of special textured paint covers 200 ft² and is on sale for $22.99/gallon. The wall needs to be painted twice (the wall needs two coats of paint). Find the cost of using paint to cover the wall.

2. A classroom has a length of 30 ft. and a width of 20 ft. The flooring is to be replaced by tiles. If each tile has a length of 36 in. and a width of 24 in., how many tiles are needed to cover the classroom floor?

3. Challenge: Assume that the tiles from Problem 2 are unavailable. Another design is available, but the tiles are square, 18 in. on a side. If these are to be installed, how many must be ordered?

4. A rectangular flower bed measures 10 m by 6 m. It has a path 2 m wide around it. Find the area of the path.

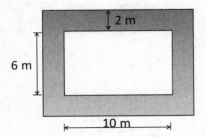

5. A diagram of Tracy's deck is shown below, shaded blue. He wants to cover the missing portion of his deck with soil in order to grow a garden.

    a. Find the area of the missing portion of the deck. Write the expression and evaluate it.

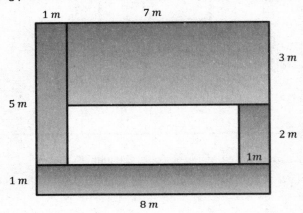

    b. Find the missing portion of the deck using a different method. Write the expression and evaluate it.

    c. Write two equivalent expressions that can be used to determine the area of the missing portion of the deck.

    d. Explain how each expression demonstrates a different understanding of the diagram.

6. The entire large rectangle below has an area of $3\frac{1}{2}$ ft². If the dimensions of the white rectangle are as shown below, write and solve an equation to find the area, A, of the shaded region.

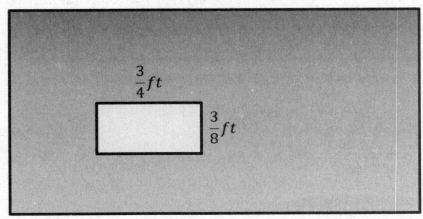

EUREKA
MATH

Example

Determine the lengths of the given line segments by determining the distance between the two endpoints.

Line Segment	Point	Point	Distance	Proof
$\overline{AB}$				
$\overline{BC}$				
$\overline{CD}$				
$\overline{BD}$				
$\overline{DE}$				
$\overline{EF}$				
$\overline{FG}$				
$\overline{EG}$				
$\boldsymbol{\overline{GA}}$				
$\overline{FA}$				
$\overline{EA}$				

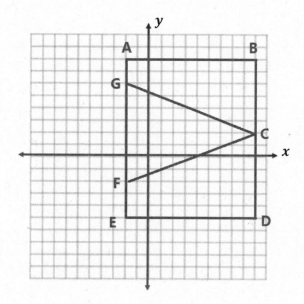

EUREKA
MATH®

Exercise

Complete the table using the diagram on the coordinate plane.

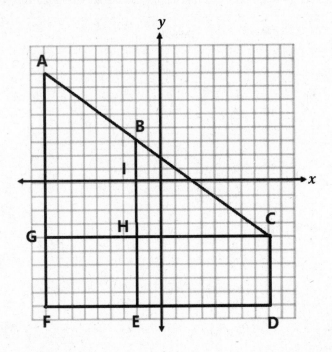

Line Segment	Point	Point	Distance	Proof
$\overline{BI}$				
$\overline{BH}$				
$\overline{BE}$				
$\overline{GH}$				
$\overline{HC}$				
$\overline{GC}$				
$\overline{CD}$				
$\overline{FG}$				
$\overline{GA}$				
$\overline{AF}$				

Lesson 7:   Distance on the Coordinate Plane

EUREKA
MATH

Extension

For each problem below, write the coordinates of two points that are 5 units apart with the segment connecting these points having the following characteristics.

    a.   The segment is vertical.

    b.   The segment intersects the x-axis.

    c.   The segment intersects the y-axis.

    d.   The segment is vertical and lies above the x-axis.

Name _____   Date _____

Use absolute value to show the lengths of *AB*, *BC*, *CD*, *DE*, and *EF*.

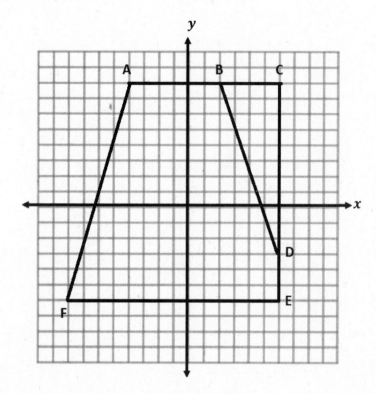

Line Segment	Point	Point	Distance	Proof
$\overline{AB}$				
$\overline{BC}$				
$\overline{CD}$				
$\overline{DE}$				
$\overline{EF}$				

EUREKA MATH®

1. Given the pairs of points, determine whether the segment that joins them will be horizontal, vertical, or neither.

   a. $X(6, 1)$ and $Y(-4.5, 1)$ _____*Horizontal*_____

   > The y-coordinates are the same, so a horizontal segment can connect these points.

   b. $M(2, -3)$ and $N(-2, 3)$ _____*Neither*_____

   > The x- and y-coordinates are different, so neither a horizontal nor a vertical line can connect these points.

   c. $E(-12, 5)$ and $F(-12, 8)$ _____*Vertical*_____

   > The x-coordinates are the same, so a vertical segment can connect these points.

2. Complete the table using absolute value to determine the lengths of the line segments.

   > I will focus on the coordinates that are different. Here, the x-coordinates are different. They have different signs, so they are on opposite sides of 0. I will add the absolute value of each coordinate to find the distance, or the length of the line segment that will connect the points.

Line Segment	Point	Point	Distance	Proof
$\overline{AB}$	$(-2, 4)$	$(5, 4)$	7	$|-2| + |5| = 7$
$\overline{CD}$	$(2, -6)$	$(2, -3)$	3	$|-6| - |-3| = 3$

   > Here, the y-coordinates are different, and they have the same sign, so they are on the same side of 0. I will subtract the absolute value of each coordinate to find the distance, or the length of the line segment that will connect the points.

3. Complete the table using the diagram and absolute value to determine the lengths of the line segments.

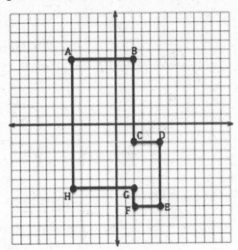

Line Segment	Point	Point	Distance	Proof
$\overline{AB}$	$(-5, 7)$	$(2, 7)$	7	$\lvert-5\rvert + \lvert2\rvert = 7$
$\overline{BC}$	$(2, 7)$	$(2, -2)$	9	$\lvert7\rvert + \lvert-2\rvert = 9$
$\overline{CD}$	$(2, -2)$	$(5, -2)$	3	$\lvert5\rvert - \lvert2\rvert = 3$
$\overline{DE}$	$(5, -2)$	$(5, -9)$	7	$\lvert-9\rvert - \lvert-2\rvert = 7$
$\overline{EF}$	$(5, -9)$	$(2, -9)$	3	$\lvert5\rvert - \lvert2\rvert = 3$
$\overline{FG}$	$(2, -9)$	$(2, -7)$	2	$\lvert-9\rvert - \lvert-7\rvert = 2$
$\overline{GH}$	$(2, -7)$	$(-5, -7)$	7	$\lvert2\rvert + \lvert-5\rvert = 7$
$\overline{HA}$	$(-5, -7)$	$(-5, 7)$	14	$\lvert-7\rvert + \lvert7\rvert = 14$

I can use absolute value just like I did in Problem 2 to determine the lengths of the line segments. I can also use the diagram in this problem to count the distance between each point to check my answer.

Lesson 7:   Distance on the Coordinate Plane

EUREKA MATH®

1. Given the pairs of points, determine whether the segment that joins them is horizontal, vertical, or neither.

   a. $X(3, 5)$ and $Y(-2, 5)$ _____

   b. $M(-4, 9)$ and $N(4, -9)$ _____

   c. $E(-7, 1)$ and $F(-7, 4)$ _____

2. Complete the table using absolute value to determine the lengths of the line segments.

Line Segment	Point	Point	Distance	Proof
$\overline{AB}$	$(-3, 5)$	$(7, 5)$		
$\overline{CD}$	$(1, -3)$	$(-6, -3)$		
$\overline{EF}$	$(2, -9)$	$(2, -3)$		
$\overline{GH}$	$(6, 1)$	$(6, 16)$		
$\overline{JK}$	$(-3, 0)$	$(-3, 12)$		

3. Complete the table using the diagram and absolute value to determine the lengths of the line segments.

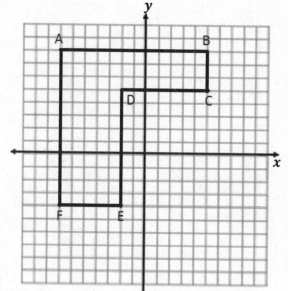

Line Segment	Point	Point	Distance	Proof
$\overline{AB}$				
$\overline{BC}$				
$\overline{CD}$				
$\overline{DE}$				
$\overline{EF}$				
$\overline{FA}$				

4. Complete the table using the diagram and absolute value to determine the lengths of the line segments.

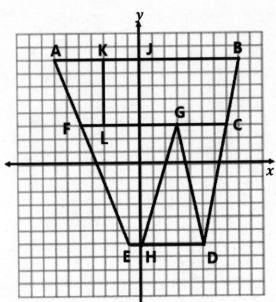

Line Segment	Point	Point	Distance	Proof
$\overline{AB}$				
$\overline{CG}$				
$\overline{CF}$				
$\overline{GF}$				
$\overline{DH}$				
$\overline{DE}$				
$\overline{HJ}$				
$\overline{KL}$				

5. Name two points in different quadrants that form a vertical line segment that is 8 units in length.

6. Name two points in the same quadrant that form a horizontal line segment that is 5 units in length.

EUREKA MATH

Examples

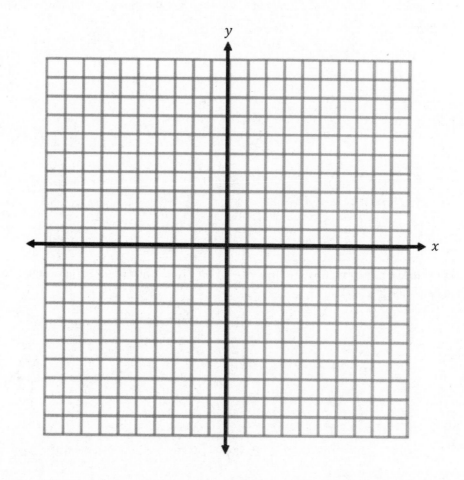

1. Plot and connect the points (3, 2), B(3, 7), and C(8, 2). Name the shape, and determine the area of the polygon.

EUREKA
MATH®

2. Plot and connect the points $E(-8, 8)$, $F(-2, 5)$, and $G(7, 2)$. Then give the best name for the polygon, and determine the area.

3. Plot and connect the following points: $K(-9, -7)$, $L(-4, -2)$, $M(-1, -5)$, and $N(-5, -5)$. Give the best name for the polygon, and determine the area.

4. Plot and connect the following points: $P(1, -4)$, $Q(5, -2)$, $R(9, -4)$, $S(7, -8)$, and $T(3, -8)$. Give the best name for the polygon, and determine the area.

Lesson 8:   Drawing Polygons in the Coordinate Plane

EUREKA MATH

5. Two of the coordinates of a rectangle are $A(3, 7)$ and $B(3, 2)$. The rectangle has an area of 30 square units. Give the possible locations of the other two vertices by identifying their coordinates. (Use the coordinate plane to draw and check your answer.)

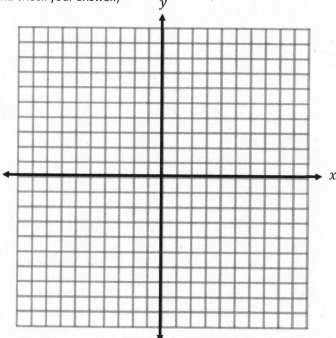

Exercises

For Exercises 1 and 2, plot the points, name the shape, and determine the area of the shape. Then write an expression that could be used to determine the area of the figure. Explain how each part of the expression corresponds to the situation.

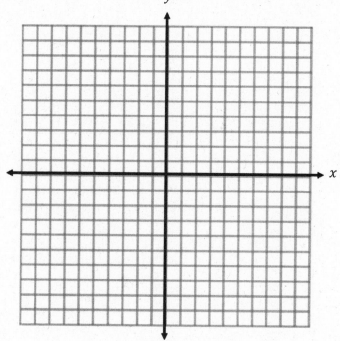

1. $A\ (4, 6)$, $B\ (8, 6)$, $C\ (10, 2)$, $D\ (8, -3)$, $E\ (5, -3)$, and $F\ (2, 2)$

2. $X\ (-9, 6)$, $Y\ (-2, -1)$, and $Z\ (-8, -7)$

Lesson 8:   Drawing Polygons in the Coordinate Plane

EUREKA MATH

3. A rectangle with vertices located at $(-3, 4)$ and $(5, 4)$ has an area of 32 square units. Determine the location of the other two vertices.

4. Challenge: A triangle with vertices located at $(-2, -3)$ and $(3, -3)$ has an area of 20 square units. Determine one possible location of the other vertex.

Name _____   Date _____

Determine the area of both polygons on the coordinate plane, and explain why you chose the methods you used. Then write an expression that could be used to determine the area of the figure. Explain how each part of the expression corresponds to the situation.

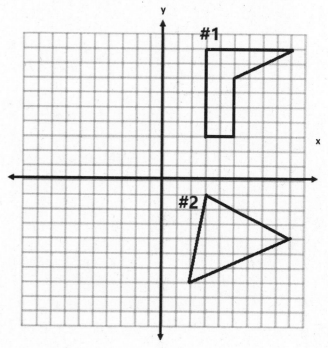

Plot the points for each shape, determine the area of the polygon, and then write an expression that could be used to determine the area of the figure. Explain how each part of the expression corresponds to the situation.

1. $A(-3, 4), B(1, -4), C(4, 2)$

    The shape formed is a triangle.

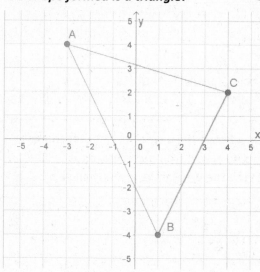

> Points are written as (x, y). So the first number tells me to move left or right from the origin, and the second number tells me to move up or down from the origin.

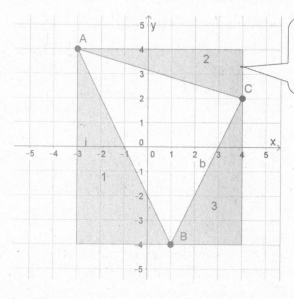

> I am unable to determine the base and height of the triangle by counting. Instead, I can draw a rectangle around the triangle.

Area of Rectangle

$A = lw$

$A = (7 \text{ units})(8 \text{ units})$

$A = 56 \text{ units}^2$

Area of Triangle 1

$A = \frac{1}{2} bh$

$A = \frac{1}{2}(8 \text{ units})(4 \text{ units})$

$A = 16 \text{ units}^2$

Area of Triangle 2

$A = \frac{1}{2} bh$

$A = \frac{1}{2}(7 \text{ units})(2 \text{ units})$

$A = 7 \text{ units}^2$

Area of Triangle 3

$A = \frac{1}{2} bh$

$A = \frac{1}{2}(6 \text{ units})(3 \text{ units})$

$A = 9 \text{ units}^2$

Total Area of Triangle

$A = 56 \text{ units}^2 - 16 \text{ units}^2 - 7 \text{ units}^2 - 9 \text{ units}^2$

$A = 24 \text{ units}^2$

> I can subtract the area of each outer triangle from the area of the rectangle, and I will be left with the area of the given triangle.

The area of the triangle is 24 units^2.

Expression: $(7)(8) - \frac{1}{2}(8)(4) - \frac{1}{2}(7)(2) - \frac{1}{2}(6)(3)$

The first term in the expression is the area of a rectangle that goes around the given triangle.

The other terms represent the area of each triangle that surrounds the given triangle.

2. $A(2, 6), B(-4, 3), C(-3, -2), D(3, -2), E(6, 3)$

    The shape formed is a pentagon.

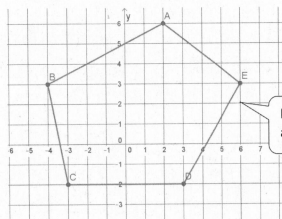

> I don't know a formula for calculating the area of a pentagon. I will need to use another method.

EUREKA MATH

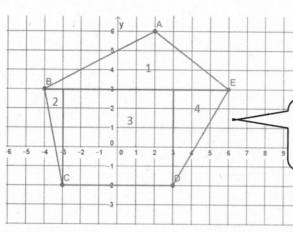

I try to decompose the shape into several smaller polygons instead of drawing the rectangle around the outside.

Area of Triangle 1

$A = \frac{1}{2}bh$

$A = \frac{1}{2}(10\,\text{units})(3\,\text{units})$

$A = 15\,\text{units}^2$

Area of Triangle 2

$A = \frac{1}{2}bh$

$A = \frac{1}{2}(1\,\text{unit})(5\,\text{units})$

$A = 2.5\,\text{units}^2$

Area of Rectangle 3

$A = lw$

$A = (6\,\text{units})(5\,\text{units})$

$A = 30\,\text{units}^2$

Area of Triangle 4

$A = \frac{1}{2}bh$

$A = \frac{1}{2}(3\,\text{units})(5\,\text{units})$

$A = 7.5\,\text{units}^2$

Total Area of Pentagon

$A = 15\,\text{units}^2 + 2.5\,\text{units}^2 + 30\,\text{units}^2 + 7.5\,\text{units}^2$

$A = 55\,\text{units}^2$

Now that I have the area of all the parts, I can add them together to get the area of the pentagon.

The area of the pentagon is $55\,\text{units}^2$.

Expression: $(6)(5) + \frac{1}{2}(10)(3) + \frac{1}{2}(1)(5) + \frac{1}{2}(3)(5)$

The first term in the expression is the area of a rectangle that is part of the pentagon. Each of the other terms represents the area of each of the triangles that make up the rest of the pentagon.

3. A triangle with vertices located at $(-5, 8)$ and $(3, 8)$ has an area of 40 square units. Determine one possible location of the other vertex.

> I know that the area of a triangle is half the area of a rectangle with the same base and height. 40 is half of 80, so the product of the base and height must be equal to 80.

One possible location of the third point would be $(-1, -2)$.

> The two given points are 8 units apart. This could be the base. Now I need a third point that is 10 units from the base so that the height will be equal to 10 units.

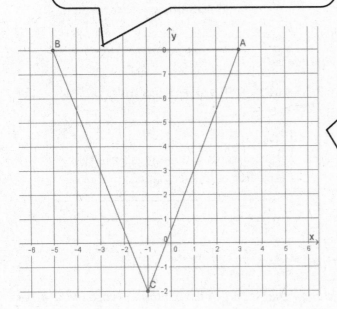

> The area of this triangle would be
>
> $$A = \frac{1}{2}bh$$
>
> $$A = \frac{1}{2}(8 \text{ units})(10 \text{ units})$$
>
> $$A = 40 \text{ units}^2$$
>
> I could have chosen any point on the line $y = -2$ to form the triangle. I could have also gone up ten units from the base instead of down to form the third point.

Lesson 8:   Drawing Polygons in the Coordinate Plane

EUREKA MATH

Plot the points for each shape, determine the area of the polygon, and then write an expression that could be used to determine the area of the figure. Explain how each part of the expression corresponds to the situation.

1. $A(1, 3)$, $B(2, 8)$, $C(8, 8)$, $D(10, 3)$, and $E(5, -2)$

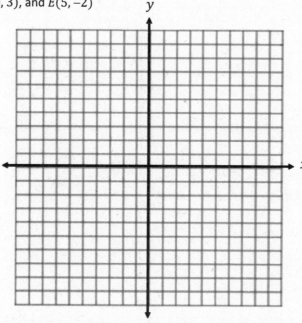

2. $X(-10, 2)$, $Y(-3, 6)$, and $Z(-6, -5)$

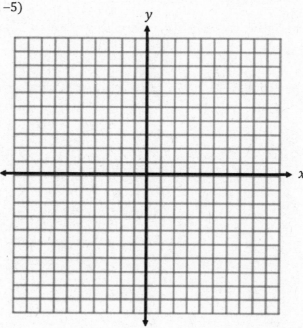

3. $E(5, 7)$, $F(9, -5)$, and $G(1, -3)$

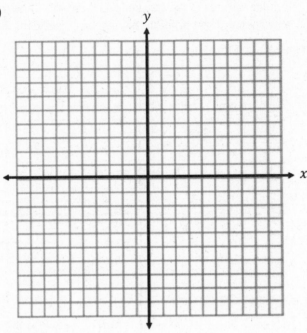

4. Find the area of the triangle in Problem 3 using a different method. Then, compare the expressions that can be used for both solutions in Problems 3 and 4.

5. Two vertices of a rectangle are $(8, -5)$ and $(8, 7)$. If the area of the rectangle is 72 square units, name the possible location of the other two vertices.

6. A triangle with two vertices located at $(5, -8)$ and $(5, 4)$ has an area of 48 square units. Determine one possible location of the other vertex.

Lesson 8:   Drawing Polygons in the Coordinate Plane

EUREKA
MATH

Number Correct: _____

Addition and Subtraction Equations—Round 1

Directions: Find the value of m in each equation.

1.	$m + 4 = 11$	
2.	$m + 2 = 5$	
3.	$m + 5 = 8$	
4.	$m - 7 = 10$	
5.	$m - 8 = 1$	
6.	$m - 4 = 2$	
7.	$m + 12 = 34$	
8.	$m + 25 = 45$	
9.	$m + 43 = 89$	
10.	$m - 20 = 31$	
11.	$m - 13 = 34$	
12.	$m - 45 = 68$	
13.	$m + 34 = 41$	
14.	$m + 29 = 52$	
15.	$m + 37 = 61$	
16.	$m - 43 = 63$	
17.	$m - 21 = 40$	

18.	$m - 54 = 37$	
19.	$4 + m = 9$	
20.	$6 + m = 13$	
21.	$2 + m = 31$	
22.	$15 = m + 11$	
23.	$24 = m + 13$	
24.	$32 = m + 28$	
25.	$4 = m - 7$	
26.	$3 = m - 5$	
27.	$12 = m - 14$	
28.	$23.6 = m - 7.1$	
29.	$14.2 = m - 33.8$	
30.	$2.5 = m - 41.8$	
31.	$64.9 = m + 23.4$	
32.	$72.2 = m + 38.7$	
33.	$1.81 = m - 15.13$	
34.	$24.68 = m - 56.82$	

EUREKA MATH®

Addition and Subtraction Equations—Round 2

Number Correct: _____

Improvement: _____

Directions: Find the value of m in each equation.

1.	$m + 2 = 7$	
2.	$m + 4 = 10$	
3.	$m + 8 = 15$	
4.	$m + 7 = 23$	
5.	$m + 12 = 16$	
6.	$m - 5 = 2$	
7.	$m - 3 = 8$	
8.	$m - 4 = 12$	
9.	$m - 14 = 45$	
10.	$m + 23 = 40$	
11.	$m + 13 = 31$	
12.	$m + 23 = 48$	
13.	$m + 38 = 52$	
14.	$m - 14 = 27$	
15.	$m - 23 = 35$	
16.	$m - 17 = 18$	
17.	$m - 64 = 1$	

18.	$6 = m + 3$	
19.	$12 = m + 7$	
20.	$24 = m + 16$	
21.	$13 = m + 9$	
22.	$32 = m - 3$	
23.	$22 = m - 12$	
24.	$34 = m - 10$	
25.	$48 = m + 29$	
26.	$21 = m + 17$	
27.	$52 = m + 37$	
28.	$\dfrac{6}{7} = m + \dfrac{4}{7}$	
29.	$\dfrac{2}{3} = m - \dfrac{5}{3}$	
30.	$\dfrac{1}{4} = m - \dfrac{8}{3}$	
31.	$\dfrac{5}{6} = m - \dfrac{7}{12}$	
32.	$\dfrac{7}{8} = m - \dfrac{5}{12}$	
33.	$\dfrac{7}{6} + m = \dfrac{16}{3}$	
34.	$\dfrac{1}{3} + m = \dfrac{13}{15}$	

Lesson 9: Determining Perimeter and Area of Polygons on the Coordinate Plane

107

Example 1

Jasjeet has made a scale drawing of a vegetable garden she plans to make in her backyard. She needs to determine the perimeter and area to know how much fencing and dirt to purchase. Determine both the perimeter and area.

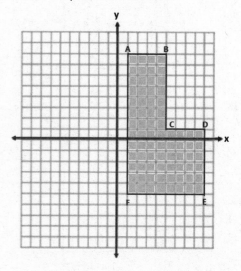

Example 2

Calculate the area of the polygon using two different methods. Write two expressions to represent the two methods, and compare the structure of the expressions.

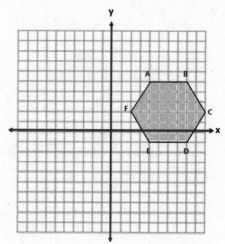

EUREKA
MATH®

Exercises

1. Determine the area of the following shapes.

   a.

   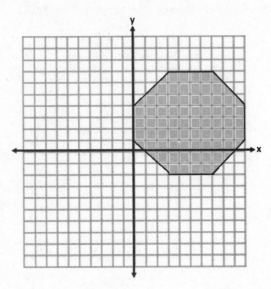

   b.

   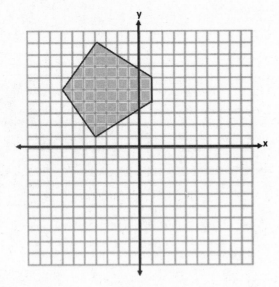

      Lesson 9:   Determining Perimeter and Area of Polygons
                               on the Coordinate Plane

EUREKA MATH®

2. Determine the area and perimeter of the following shapes.

a.

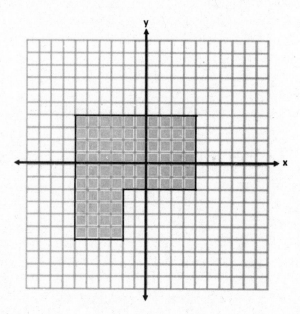

b.

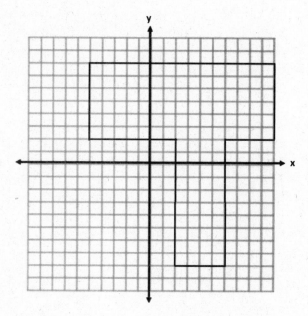

Name _____   Date _____

Determine the area and perimeter of the figure below. Note that each square unit is 1 unit in length.

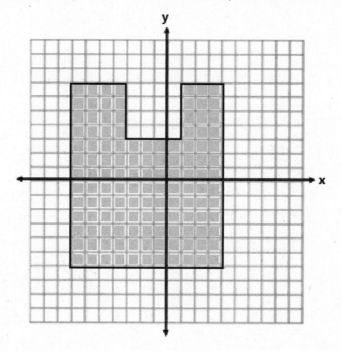

Use the diagram to answer the following questions.

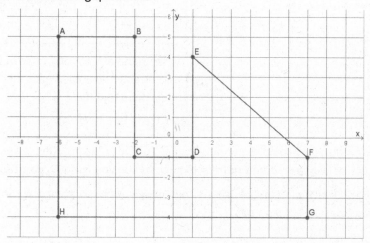

a. Determine the area of the polygon.

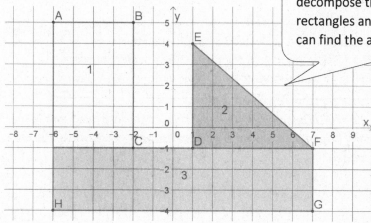

Similar to Lesson 8, I can decompose the polygon into rectangles and triangles. Then I can find the area of each piece.

Area of Rectangle 1

$A = lw$

$A = (4 \text{ units})(6 \text{ units})$

$A = 24 \text{ units}^2$

Area of Triangle 2

$A = \frac{1}{2}bh$

$A = \frac{1}{2}(6 \text{ units})(5 \text{ units})$

$A = 15 \text{ units}^2$

Area of Rectangle 3

$A = lw$

$A = (13 \text{ units})(3 \text{ units})$

$A = 39 \text{ units}^2$

Total Area of Polygon

$A = 24 \text{ units}^2 + 15 \text{ units}^2 + 39 \text{ units}^2$

$A = 78 \text{ units}^2$

b. Write an expression that could be used to determine the area.

$$(4)(6) + \frac{1}{2}(6)(5) + (13)(3)$$

> I write the expression by showing how I determine the area of each piece and then add them together.

c. Describe another method you could use to find the area of the polygon. Then, state how the expression for the area would be different from the expression you wrote in part (b).

Instead, I could have drawn a rectangle around the outside of the polygon and subtracted the area of the extra pieces from the total area.

d. If the length of each square was worth 5 units instead of 1 unit, how would the area of polygon change? How would your expression change to represent this area?

$5l \times 5w = 25lw$

> If each square is 5 units instead of 1 unit, then the length would be 5 times as long as the original length, and the width would be 5 times as long as the original width.

The area will be 25 times larger than the original area when the side lengths are five times longer than the original lengths. Therefore, I could multiply my entire expression by 25 to make the area of the original polygon 25 times bigger.

$$25\left[(4)(6) + \frac{1}{2}(6)(5) + (13)(3)\right]$$

EUREKA MATH

1. Determine the area of the polygon.

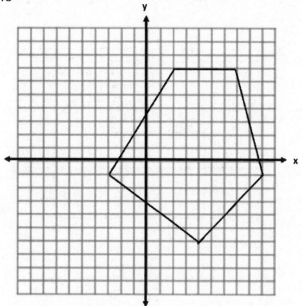

2. Determine the area and perimeter of the polygon.

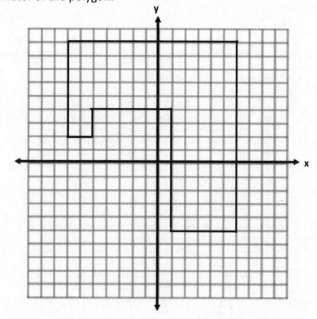

EUREKA
MATH®

Lesson 9:   Determining Perimeter and Area of Polygons
on the Coordinate Plane

© 2019 Great Minds®. eureka-math.org

117

3. Determine the area of the polygon. Then, write an expression that could be used to determine the area.

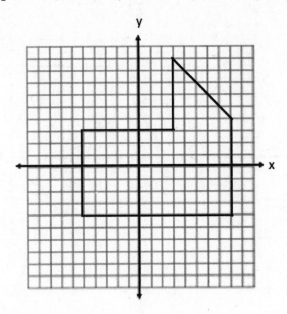

4. If the length of each square was worth 2 instead of 1, how would the area in Problem 3 change? How would your expression change to represent this area?

5. Determine the area of the polygon. Then, write an expression that represents the area.

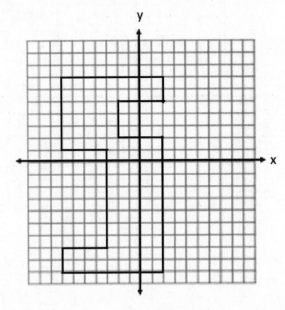

6. Describe another method you could use to find the area of the polygon in Problem 5. Then, state how the expression for the area would be different than the expression you wrote.

EUREKA
MATH®

7. Write one of the letters from your name using rectangles on the coordinate plane. Then, determine the area and perimeter. (For help see Exercise 2(b). This irregular polygon looks sort of like a T.)

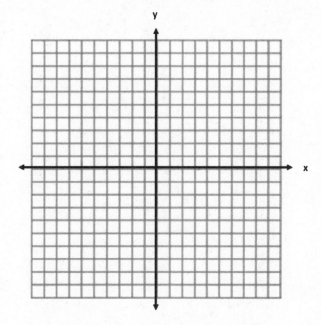

Opening Exercise

a. Find the area and perimeter of this rectangle:

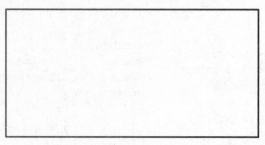

5 cm

9 cm

b. Find the width of this rectangle. The area is 1.2 m², and the length is 1.5 m.

$A = 1.2\ m^2$ $w = ?$

$l = 1.5\ m$

Example: Student Desks or Tables

1. Measure the dimensions of the top of your desk.

2. How do you find the area of the top of your desk?

3. How do you find the perimeter?

4. Record these on your paper in the appropriate column.

Exploratory Challenge

Estimate and predict the area and perimeter of each object. Then measure each object, and calculate both the area and perimeter of each.

Object or Item to be Measured	Measurement Units	Precision (measure to the nearest)	Area Prediction (square units)	Area (square units) Write the expression and evaluate it.	Perimeter Prediction (linear units)	Perimeter (linear units)
Ex: door	feet	half foot		$6\frac{1}{2}\text{ft.} \times 3\frac{1}{2}\text{ft.}$ $= 22\frac{3}{4}\text{ft}^2$		$2\left(3\frac{1}{2}\text{ft.} + 6\frac{1}{2}\text{ft.}\right)$ $= 20\text{ ft.}$
desktop						

Lesson 10:   Distance, Perimeter, and Area in the Real World

EUREKA MATH

Optional Challenge

Object or Item to be Measured	Measurement Units	Precision (measure to the nearest)	Area (square units)	Perimeter (linear units)
Ex: door	feet	half foot	$6\frac{1}{2}$ ft. $\times 3\frac{1}{2}$ ft. $= 22\frac{3}{4}$ ft²	$2\left(3\frac{1}{2}\text{ ft.} + 6\frac{1}{2}\text{ ft.}\right)$ $= 20$ ft.

Name _____ Date _____

1. The local school is building a new playground. This plan shows the part of the playground that needs to be framed with wood for the swing set. The unit of measure is feet. Determine the number of feet of wood needed to fram the area.

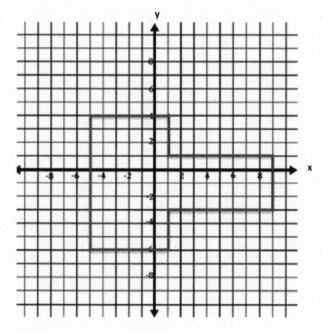

2. The school wants to fill the area enclosed with wood with mulch for safety. Determine the area in square feet that needs to be covered by the mulch.

1. Lukas and Juan are modeling rectangular designs using toothpicks. They have been given 30 toothpicks. They agree that they should only use whole toothpicks and not break any apart.

> I could draw pictures of possible rectangles might help get me started.

a. What are all of the possible dimensions of the rectangular designs?

Length (in toothpicks)	Width (in toothpicks)
14	1
13	2
12	3
11	4
10	5
9	6
8	7

> There are two lengths and two widths in each rectangle. So I want half of the 30 toothpicks, 15, used in one length and one width.

b. Which rectangular design yields a maximum area? Which design yields the minimum area?

Length (in toothpicks)	Width (in toothpicks)	Area (in toothpicks squared)
14	1	14
13	2	26
12	3	36
11	4	44
10	5	50
9	6	54
8	7	56

I can create a table of all the possible areas so that I can compare them to answer the question.

The 8 toothpicks by 7 toothpicks design would have the maximum area of 56 toothpicks squared, while the 14 toothpicks by 1 toothpick design would only have an area of 14 toothpicks squared.

2. Maria is designing a new rectangular block for a quilt.

I need to include all of the lines where Maria would be sewing, not just the outside edges. I might want to label some of the sides where the measurements are not shown.

10 in.
13 in.
13 in.
12 in.

a. Maria must sew along all of the edges. Determine the total length that must be sewn.

10 in. + 10 in. + 13 in. + 13 in. + 12 in. + 12 in. = 70 in.

Maria must sew across 70 inches.

b. If Maria's spool contains 110 yards (3,960 inches) of thread, how many complete blocks could she sew?

3,960 in. ÷ 70 in. per block ≈ 56.57 blocks

Maria would be able to complete 56 blocks.

EUREKA MATH

3. This is a drawing of the country of Sweden's flag. The area of the flag is $30\frac{5}{8}$ ft².

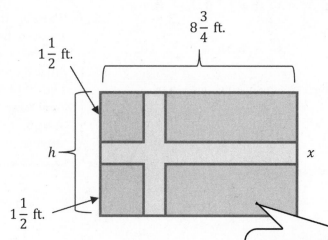

$8\frac{3}{4}$ ft.

$1\frac{1}{2}$ ft.

h

$1\frac{1}{2}$ ft.

x

> I have been given the area of the rectangle. So I can work backward to determine the height.

a. Determine the value of the height of the flag.

$$A = bh$$

$$h = A \div b$$

$$h = 30\frac{5}{8}\,\text{ft}^2 \div 8\frac{3}{4}\,\text{ft.}$$

$$h = \frac{245}{8}\,\text{ft}^2 \div \frac{35}{4}\,\text{ft.}$$

$$h = \frac{7}{2}\,\text{ft.}$$

$$h = 3\frac{1}{2}\,\text{ft.}$$

The height of the flag is $3\frac{1}{2}$ feet.

EUREKA
MATH®

b. Using what you found in part (a), determine the missing value of the height.

$$3\frac{1}{2}\text{ ft.} = 1\frac{1}{2}\text{ ft.} + 1\frac{1}{2}\text{ ft.} + x$$

$$3\frac{1}{2}\text{ ft.} = 3\text{ ft.} + x$$

$$3\frac{1}{2}\text{ ft.} - 3\text{ ft.} = 3\text{ ft.} + x - 3\text{ ft.}$$

$$\frac{1}{2}\text{ ft.} = x$$

There are three portions that make up the height of the flag. I know the lengths of two of them and the total. I can use this to solve for the length of the unknown piece.

The length of the missing piece is $\frac{1}{2}$ foot.

EUREKA MATH

1. How is the length of the side of a square related to its area and perimeter? The diagram below shows the first four squares stacked on top of each other with their upper left-hand corners lined up. The length of one side of the smallest square is 1 foot.

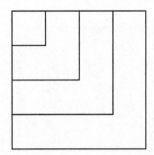

a. Complete this chart calculating area and perimeter for each square.

Side Length (in feet)	Expression Showing the Area	Area (in square feet)	Expression Showing the Perimeter	Perimeter (in feet)
1	1×1	1	1×4	4
2				
3				
4				
5				
6				
7				
8				
9				
10				
n				

b. In a square, which numerical value is greater, the area or the perimeter?

c. When is the numerical value of a square's area (in square units) equal to its perimeter (in units)?

d. Why is this true?

2. This drawing shows a school pool. The walkway around the pool needs special nonskid strips installed but only at the edge of the pool and the outer edges of the walkway.

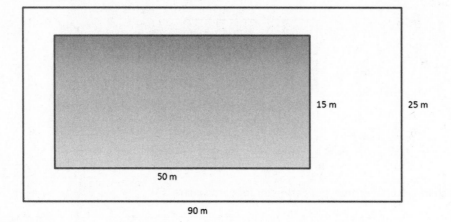

a. Find the length of nonskid strips that is needed for the job.

b. The nonskid strips are sold only in rolls of 50 m. How many rolls need to be purchased for the job?

3. A homeowner called in a painter to paint the walls and ceiling of one bedroom. His bedroom is 18 ft. long, 12 ft. wide, and 8 ft. high. The room has <u>two</u> doors, each 3 ft. by 7 ft., and <u>three</u> windows each 3 ft. by 5 ft. The doors and windows will not be painted. A gallon of paint can cover 300 ft^2. A hired painter claims he needs a minimum of 4 gallons. Show that his estimate is too high.

4. Theresa won a gardening contest and was awarded a roll of deer-proof fencing. The fencing is 36 feet long. She and her husband, John, discuss how to best use the fencing to make a rectangular garden. They agree that they should only use whole numbers of feet for the length and width of the garden.

a. What are all of the possible dimensions of the garden?

b. Which plan yields the maximum area for the garden? Which plan yields the minimum area?

5. Write and then solve the equation to find the missing value below.

$$A = 1.82 \; m^2 \qquad w = ?$$

$$l = 1.4 \; m$$

EUREKA MATH

6.   Challenge: This is a drawing of the flag of the Republic of the Congo. The area of this flag is $3\frac{3}{4}$ ft².

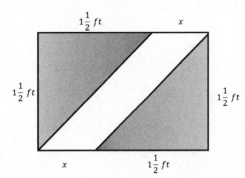

a.   Using the area formula, tell how you would determine the value of the base. This figure is not drawn to scale.

b.   Using what you found in part (a), determine the missing value of the base.

Number Correct: _____

Multiplication of Fractions II—Round 1

Directions: Determine the product of the fractions and simplify.

1.	$\frac{1}{2} \times \frac{5}{8}$		16.	$\frac{2}{9} \times \frac{3}{8}$	
2.	$\frac{3}{4} \times \frac{3}{5}$		17.	$\frac{3}{8} \times \frac{8}{9}$	
3.	$\frac{1}{4} \times \frac{7}{8}$		18.	$\frac{3}{4} \times \frac{7}{9}$	
4.	$\frac{3}{9} \times \frac{2}{5}$		19.	$\frac{3}{5} \times \frac{10}{13}$	
5.	$\frac{5}{8} \times \frac{3}{7}$		20.	$1\frac{2}{7} \times \frac{7}{8}$	
6.	$\frac{3}{7} \times \frac{4}{9}$		21.	$3\frac{1}{2} \times 3\frac{5}{6}$	
7.	$\frac{2}{5} \times \frac{3}{8}$		22.	$1\frac{7}{8} \times 5\frac{1}{5}$	
8.	$\frac{4}{9} \times \frac{5}{9}$		23.	$5\frac{4}{5} \times 3\frac{2}{9}$	
9.	$\frac{2}{3} \times \frac{5}{7}$		24.	$7\frac{2}{5} \times 2\frac{3}{8}$	
10.	$\frac{2}{7} \times \frac{3}{10}$		25.	$4\frac{2}{3} \times 2\frac{3}{10}$	
11.	$\frac{3}{4} \times \frac{9}{10}$		26.	$3\frac{3}{5} \times 6\frac{1}{4}$	
12.	$\frac{3}{5} \times \frac{2}{9}$		27.	$2\frac{7}{9} \times 5\frac{1}{3}$	
13.	$\frac{2}{10} \times \frac{5}{6}$		28.	$4\frac{3}{8} \times 3\frac{1}{5}$	
14.	$\frac{5}{8} \times \frac{7}{10}$		29.	$3\frac{1}{3} \times 5\frac{2}{5}$	
15.	$\frac{3}{5} \times \frac{7}{9}$		30.	$2\frac{2}{3} \times 7$	

Number Correct: _____

Improvement: _____

Multiplication of Fractions II—Round 2

Directions: Determine the product of the fractions and simplify.

1.	$\dfrac{2}{3} \times \dfrac{5}{7}$	
2.	$\dfrac{1}{4} \times \dfrac{3}{5}$	
3.	$\dfrac{2}{3} \times \dfrac{2}{5}$	
4.	$\dfrac{5}{9} \times \dfrac{5}{8}$	
5.	$\dfrac{5}{8} \times \dfrac{3}{7}$	
6.	$\dfrac{3}{4} \times \dfrac{7}{8}$	
7.	$\dfrac{2}{5} \times \dfrac{3}{8}$	
8.	$\dfrac{3}{4} \times \dfrac{3}{4}$	
9.	$\dfrac{7}{8} \times \dfrac{3}{10}$	
10.	$\dfrac{4}{9} \times \dfrac{1}{2}$	
11.	$\dfrac{6}{11} \times \dfrac{3}{8}$	
12.	$\dfrac{5}{6} \times \dfrac{9}{10}$	
13.	$\dfrac{3}{4} \times \dfrac{2}{9}$	
14.	$\dfrac{4}{11} \times \dfrac{5}{8}$	
15.	$\dfrac{2}{3} \times \dfrac{9}{10}$	

16.	$\dfrac{3}{11} \times \dfrac{2}{9}$	
17.	$\dfrac{3}{5} \times \dfrac{10}{21}$	
18.	$\dfrac{4}{9} \times \dfrac{3}{10}$	
19.	$\dfrac{3}{8} \times \dfrac{4}{5}$	
20.	$\dfrac{6}{11} \times \dfrac{2}{15}$	
21.	$1\dfrac{2}{3} \times \dfrac{3}{5}$	
22.	$2\dfrac{1}{6} \times \dfrac{3}{4}$	
23.	$1\dfrac{2}{5} \times 3\dfrac{2}{3}$	
24.	$4\dfrac{2}{3} \times 1\dfrac{1}{4}$	
25.	$3\dfrac{1}{2} \times 2\dfrac{4}{5}$	
26.	$3 \times 5\dfrac{3}{4}$	
27.	$1\dfrac{2}{3} \times 3\dfrac{1}{4}$	
28.	$2\dfrac{3}{5} \times 3$	
29.	$1\dfrac{5}{7} \times 3\dfrac{1}{2}$	
30.	$3\dfrac{1}{3} \times 1\dfrac{9}{10}$	

Opening Exercise

Which prism holds more 1 in. × 1 in. × 1 in. cubes? How many more cubes does the prism hold?

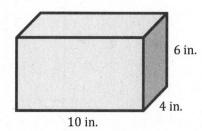 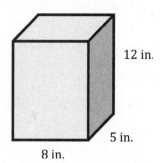

Example 1

A box with the same dimensions as the prism in the Opening Exercise is used to ship miniature dice whose side lengths have been cut in half. The dice are $\frac{1}{2}$ in. × $\frac{1}{2}$ in. × $\frac{1}{2}$ in. cubes. How many dice of this size can fit in the box?

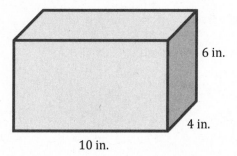

Example 2

A $\frac{1}{4}$ in. cube was used to fill the prism.

How many $\frac{1}{4}$ in. cubes does it take to fill the prism?

What is the volume of the prism?

How is the number of cubes related to the volume?

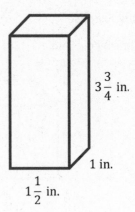

$3\frac{3}{4}$ in.

1 in.

$1\frac{1}{2}$ in.

Exercises

1. Use the prism to answer the following questions.

    a. Calculate the volume.

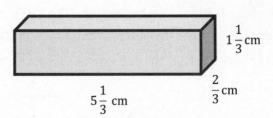

$1\frac{1}{3}$ cm

$\frac{2}{3}$ cm

$5\frac{1}{3}$ cm

    b. If you have to fill the prism with cubes whose side lengths are less than 1 cm, what size would be best?

    c. How many of the cubes would fit in the prism?

    d. Use the relationship between the number of cubes and the volume to prove that your volume calculation is correct.

EUREKA MATH

2. Calculate the volume of the following rectangular prisms.

a.

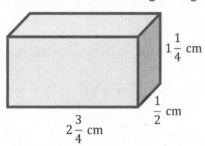

$1\frac{1}{4}$ cm

$\frac{1}{2}$ cm

$2\frac{3}{4}$ cm

b.

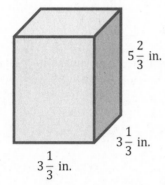

$5\frac{2}{3}$ in.

$3\frac{1}{3}$ in.

$3\frac{1}{3}$ in.

3. A toy company is packaging its toys to be shipped. Each small toy is placed inside a cube-shaped box with side lengths of $\frac{1}{2}$ in. These smaller boxes are then placed into a larger box with dimensions of

12 in. $\times\ 4\frac{1}{2}$ in. $\times\ 3\frac{1}{2}$ in.

a. What is the greatest number of small toy boxes that can be packed into the larger box for shipping?

b. Use the number of small toy boxes that can be shipped in the larger box to help determine the volume of the shipping box.

4. A rectangular prism with a volume of 8 cubic units is filled with cubes twice: once with cubes with side lengths of $\frac{1}{2}$ unit and once with cubes with side lengths of $\frac{1}{3}$ unit.

    a. How many more of the cubes with $\frac{1}{3}$-unit side lengths than cubes with $\frac{1}{2}$-unit side lengths are needed to fill the prism?

    b. Why does it take more cubes with $\frac{1}{3}$-unit side lengths to fill the prism than it does with cubes with $\frac{1}{2}$-unit side lengths?

5. Calculate the volume of the rectangular prism. Show two different methods for determining the volume.

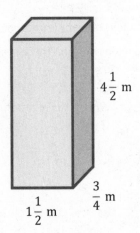

$4\frac{1}{2}$ m

$\frac{3}{4}$ m

$1\frac{1}{2}$ m

EUREKA
MATH®

Name _____   Date _____

Calculate the volume of the rectangular prism using two different methods. Label your solutions Method 1 and Method 2.

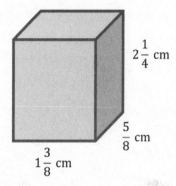

$2\frac{1}{4}$ cm

$\frac{5}{8}$ cm

$1\frac{3}{8}$ cm

1. Answer the following questions using this rectangular prism.

    a. What is the volume of the prism?

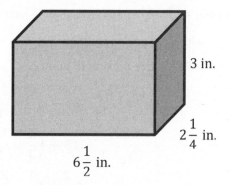

3 in.

$2\frac{1}{4}$ in.

$6\frac{1}{2}$ in.

$$V = l\,w\,h$$

$$V = \left(6\frac{1}{2}\text{ in.}\right)\left(2\frac{1}{4}\text{ in.}\right)(3\text{ in.})$$

$$V = 43\frac{7}{8}\text{ in}^3$$

The volume of the prism is $43\frac{7}{8}$ cubic inches.

    b. How many cubes with side lengths of $\frac{1}{4}$ in. would fill the prism?

> I can determine how many cubes fit along the length by finding equivalent fractions.

Cubes that fit along the length: $6\frac{1}{2} = \frac{13}{2} = \frac{26}{4}$

Cubes that fit along the width: $2\frac{1}{4} = \frac{9}{4}$

Cubes that fit along the height: $3 = \frac{12}{4}$

$$26 \times 9 \times 12 = 2{,}808$$

The prism could be filled with 2,808 cubes with a side length of $\frac{1}{4}$ in.

> I need to convert the units to fourths to find how many $\frac{1}{4}$-inch cubes fit along the length. $\frac{26}{4} = 26$ fourths, so I know 26 $\frac{1}{4}$-inch cubes fit along the length.

    c. How many cubes with side lengths of $\frac{1}{8}$ in. would fill the prism?

Number of cubes $= 2(26) \times 2(9) \times 2(12)$

$$= 52 \times 18 \times 24$$

$$= 22{,}464$$

The prism could be filled with 22,464 cubes with a side length of $\frac{1}{8}$ in.

> I know that there are two $\frac{1}{8}$ inches in every $\frac{1}{4}$ inch, which means that there will be twice the amount of cubes for each dimension.

2. A company is packaging its décor for living rooms to be shipped to stores. Some of the décor is placed inside a cube-shaped package with side lengths of $4\frac{1}{3}$ in. These packages are then placed into a shipping box with dimensions of 26 in. $\times 13$ in. $\times 8\frac{2}{3}$ in.

a. What is the maximum number of smaller packages that can be placed into the larger shipping box?

$$26 \div 4\frac{1}{3} = \frac{26}{1} \div \frac{13}{3} = \frac{26}{1} \times \frac{3}{13} = 6$$

$$13 \div 4\frac{1}{3} = \frac{13}{1} \div \frac{13}{3} = \frac{13}{1} \times \frac{3}{13} = 3$$

$$8\frac{2}{3} \div 4\frac{1}{3} = \frac{26}{3} \div \frac{13}{3} = \frac{26}{3} \times \frac{3}{13} = 2$$

I need to determine how many times $4\frac{1}{3}$ in. will fit into each dimension.

$6 \times 3 \times 2 = 36$

36 smaller packages can fit in the larger shipping box.

b. Use the number of packages that can be shipped in the box to help determine the volume of the shipping box.

$$4\frac{1}{3} \text{ in.} \times 4\frac{1}{3} \text{ in.} \times 4\frac{1}{3} \text{ in.}$$

$$\frac{13}{3} \text{ in.} \times \frac{13}{3} \text{ in.} \times \frac{13}{3} \text{ in.}$$

$$\frac{2,197}{27} \text{ in}^3$$

$$81\frac{10}{27} \text{ in}^3$$

I can change the mixed numbers to fractions greater than one before I multiply.

One small package would have a volume of $81\frac{10}{27}$ in³.

$$36 \times 81\frac{10}{27} \text{ in}^3$$

$$\frac{36}{1} \times \frac{2,197}{27} \text{ in}^3$$

$$\frac{79,092}{27} \text{ in}^3$$

$$2,929\frac{1}{3} \text{ in}^3$$

I multiply the number of cubes by the volume of the cube.

The volume of the shipping box is $2,929\frac{1}{3}$ in³.

EUREKA MATH

3. A rectangular prism has a volume of 141.599 cubic centimeters. The height of the box is 8.9 centimeters, and the width is 3.7 centimeters.

    a. Write an equation that relates the volume to the length, width, and height. Let l represent the length, in centimeters.

$$141.599 = l(3.7)(8.9)$$

> I can start with the formula $V = l\,w\,h$ and then fill in what I know.

    b. Solve the equation.

$$141.599 = 32.93l$$
$$141.599 \div 32.93 = 32.93l \div 32.93$$
$$4.3 = l$$

> This looks a lot like the equations I worked with in Module 4. I can use inverse operations to solve.

The length is 4.3 cm.

1. Answer the following questions using this rectangular prism:

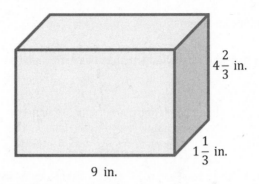

    a. What is the volume of the prism?

    b. Linda fills the rectangular prism with cubes that have side lengths of $\frac{1}{3}$ in. How many cubes does she need to fill the rectangular prism?

    c. How is the number of cubes related to the volume?

    d. Why is the number of cubes needed different from the volume?

    e. Should Linda try to fill this rectangular prism with cubes that are $\frac{1}{2}$ in. long on each side? Why or why not?

2. Calculate the volume of the following prisms.

    a.

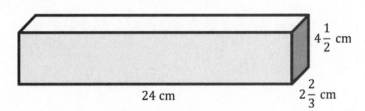

    b.

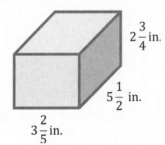

---

3. A rectangular prism with a volume of 12 cubic units is filled with cubes twice: once with cubes with $\frac{1}{2}$-unit side lengths and once with cubes with $\frac{1}{3}$-unit side lengths.

    a. How many more of the cubes with $\frac{1}{3}$-unit side lengths than cubes with $\frac{1}{2}$-unit side lengths are needed to fill the prism?

    b. Finally, the prism is filled with cubes whose side lengths are $\frac{1}{4}$ unit. How many $\frac{1}{4}$-unit cubes would it take to fill the prism?

4. A toy company is packaging its toys to be shipped. Each toy is placed inside a cube-shaped box with side lengths of $3\frac{1}{2}$ in. These smaller boxes are then packed into a larger box with dimensions of 14 in. $\times$ 7 in. $\times$ $3\frac{1}{2}$ in.

    a. What is the greatest number of toy boxes that can be packed into the larger box for shipping?

    b. Use the number of toy boxes that can be shipped in the large box to determine the volume of the shipping box.

5. A rectangular prism has a volume of 34.224 cubic meters. The height of the box is 3.1 meters, and the length is 2.4 meters.

    a. Write an equation that relates the volume to the length, width, and height. Let w represent the width, in meters.

    b. Solve the equation.

EUREKA MATH

Example 1

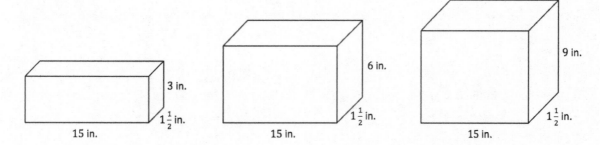

a.   Write a numerical expression for the volume of each of the rectangular prisms above.

b.   What do all of these expressions have in common? What do they represent?

c.   Rewrite the numerical expressions to show what they have in common.

d.   If we know volume for a rectangular prism as length times width times height, what is another formula for volume that we could use based on these examples?

e.   What is the area of the base for all of the rectangular prisms?

f.   Determine the volume of each rectangular prism using either method.

g.   How do the volumes of the first and second rectangular prisms compare? The volumes of the first and third?

Example 2

The base of a rectangular prism has an area of $3\frac{1}{4}$ in^2. The height of the prism is $2\frac{1}{2}$ in. Determine the volume of the rectangular prism.

Extension

A company is creating a rectangular prism that must have a volume of 6 ft^3. The company also knows that the area of the base must be $2\frac{1}{2}$ ft^2. How can you use what you learned today about volume to determine the height of the rectangular prism?

Name _____   Date _____

1. Determine the volume of the rectangular prism in two different ways.

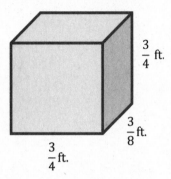

$\frac{3}{4}$ ft.

$\frac{3}{8}$ ft.

$\frac{3}{4}$ ft.

2. The area of the base of a rectangular prism is 12 cm², and the height is $3\frac{1}{3}$ cm. Determine the volume of the rectangular prism.

1. The area of the base of a rectangular prism is $7\frac{1}{2}$ m^2, and the height is $3\frac{3}{4}$ m. Determine the volume of the rectangular prism.

   V = Area of base × height

   $V = \left(7\dfrac{1}{2}\,\text{m}^2 \right)\left(3\dfrac{3}{4}\,\text{m} \right)$

   $V = \left(\dfrac{15}{2}\,\text{m}^2 \right)\left(\dfrac{15}{4}\,\text{m} \right)$

   $V = \dfrac{225}{8}\,\text{m}^3$

   $V = 28\dfrac{1}{8}\,\text{m}^3$

   > I was given the area of the base and the height. So instead of using the formula $V = l\,w\,h$, I will multiply the area of the base times the height.

2. The length of a rectangular prism is $4\frac{1}{2}$ times as long as the height. The width is $\frac{2}{3}$ of the height. The height is 12 cm. Determine the volume of the rectangular prism.

   Height $= 12\,\text{cm}$

   Length $= 12\,\text{cm} \times 4\dfrac{1}{2} = 54\,\text{cm}$

   Width $= 12\,\text{cm} \times \dfrac{2}{3} = 8\,\text{cm}$

   $V = l\,w\,h$

   $V = (54\,\text{cm})(8\,\text{cm})(12\,\text{cm})$

   $V = 5,184\,\text{cm}^3$

   The volume of the rectangular prism is 5,184 cm³.

   > Before I can calculate the volume, I need to determine the length and width of the prism, and then I can use these measurements in the formula.

3.

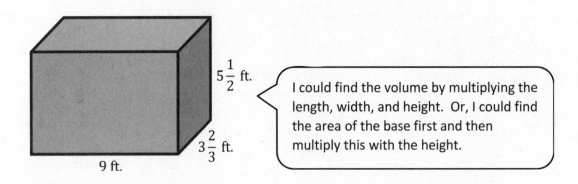

a. Write numerical expressions to represent the volume in two different ways, and explain what each reveals.

The first expression shows how I can find the volume using the formula $V = l\,w\,h$.

$$(9 \text{ ft.}) \left(3\frac{2}{3} \text{ ft.} \right) \left(5\frac{1}{2} \text{ ft.} \right)$$

The second expression shows how I can find the volume using the formula $V = B\,h$.

$$(33 \text{ ft}^2) \left(5\frac{1}{2} \text{ ft.} \right)$$

b. Determine the volume of the rectangular prism.

$$(9 \text{ ft.}) \left(3\frac{2}{3} \text{ ft.} \right) \left(5\frac{1}{2} \text{ ft.} \right) = 181\frac{1}{2} \text{ ft}^3$$

OR

$$\left(33 \text{ ft}^2 \right) \left(5\frac{1}{2} \text{ ft.} \right) = 181\frac{1}{2} \text{ ft}^3$$

EUREKA MATH®

4. The area of the base in this rectangular prism is fixed at 45 m². This means that for the varying heights, there will be various volumes.

> This means that the length and width will stay the same and only the height changes. Since the height is changing, the volume will change.

a. Complete the table of values to determine the various heights and volumes.

> I know the area of the base, so I can multiply what I know by the height to determine the volume.

Height in Meters	Volume in Cubic Meters
1	45
2	90
3	135
4	180
5	225
6	270

b. Write an equation to represent the relationship in the table. Be sure to define the variables used in the equation.

   Let x represent the height of the rectangular prism in meters.

   Let y represent the volume of the rectangular prism in cubic meters.

   $y = 45x$

> I know that the volume is equal to the area of the base times the height.

c. What is the unit rate for this relationship? What does it mean in this situation?

   The unit rate is 45.

   For every meter of height, the volume increases by 45 m² because the area of the base is 45 m². In order to determine the volume, multiply the height by 45 m².

1. Determine the volume of the rectangular prism.

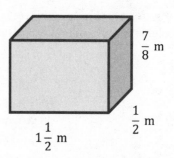

$\frac{7}{8}$ m

$\frac{1}{2}$ m

$1\frac{1}{2}$ m

2. The area of the base of a rectangular prism is $4\frac{3}{4}$ ft^2, and the height is $2\frac{1}{3}$ ft. Determine the volume of the rectangular prism.

3. The length of a rectangular prism is $3\frac{1}{2}$ times as long as the width. The height is $\frac{1}{4}$ of the width. The width is 3 cm. Determine the volume.

4.

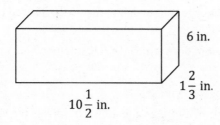

6 in.

$1\frac{2}{3}$ in.

$10\frac{1}{2}$ in.

   a. Write numerical expressions to represent the volume in two different ways, and explain what each reveals.
   b. Determine the volume of the rectangular prism.

5. An aquarium in the shape of a rectangular prism has the following dimensions: length = 50 cm, width = $25\frac{1}{2}$ cm, and height = $30\frac{1}{2}$ cm.
   a. Write numerical expressions to represent the volume in two different ways, and explain what each reveals.
   b. Determine the volume of the rectangular prism.

6. The area of the base in this rectangular prism is fixed at 36 cm². As the height of the rectangular prism changes, the volume will also change as a result.

   a. Complete the table of values to determine the various heights and volumes.

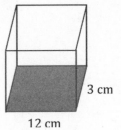

3 cm

12 cm

Height of Prism (in centimeters)	Volume of Prism (in cubic centimeters)
2	72
3	108
	144
	180
6	
7	
	288

   b. Write an equation to represent the relationship in the table. Be sure to define the variables used in the equation.

   c. What is the unit rate for this proportional relationship? What does it mean in this situation?

7. The volume of a rectangular prism is 16.328 cm³. The height is 3.14 cm.

   a. Let B represent the area of the base of the rectangular prism. Write an equation that relates the volume, the area of the base, and the height.

   b. Solve the equation for B.

EUREKA
MATH

Example 1

Determine the volume of a cube with side lengths of $2\frac{1}{4}$ cm.

Example 2

Determine the volume of a rectangular prism with a base area of $\frac{7}{12}$ ft^2 and a height of $\frac{1}{3}$ ft.

Exercises

1. Use the rectangular prism to answer the next set of questions.
   a. Determine the volume of the prism.

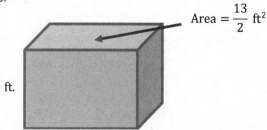

Area $= \dfrac{13}{2}$ ft^2

$\dfrac{5}{3}$ ft.

   b. Determine the volume of the prism if the height of the prism is doubled.

c. Compare the volume of the rectangular prism in part (a) with the volume of the prism in part (b). What do you notice?

d. Complete and use the table below to determine the relationship between the height and volume.

Height of Prism (in feet)	Volume of Prism (in cubic feet)
$\dfrac{5}{3}$	$\dfrac{65}{6}$
$\dfrac{10}{3}$	$\dfrac{130}{6}$
$\dfrac{15}{3}$	
$\dfrac{20}{3}$	

What happened to the volume when the height was tripled?

What happened to the volume when the height was quadrupled?

What conclusions can you make when the base area stays constant and only the height changes?

2.

a. If B represents the area of the base and h represents the height, write an expression that represents the volume.

Lesson 13:   The Formulas for Volume

EUREKA MATH

b.   If we double the height, write an expression for the new height.

c.   Write an expression that represents the volume with the doubled height.

d.   Write an equivalent expression using the commutative and associative properties to show the volume is twice the original volume.

3.   Use the cube to answer the following questions.
    a.   Determine the volume of the cube.

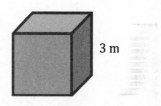

3 m

    b.   Determine the volume of a cube whose side lengths are half as long as the side lengths of the original cube.

    c.   Determine the volume if the side lengths are one-fourth as long as the original cube's side lengths.

    d.   Determine the volume if the side lengths are one-sixth as long as the original cube's side lengths.

e. Explain the relationship between the side lengths and the volumes of the cubes.

4. Check to see if the relationship you found in Exercise 3 is the same for rectangular prisms.

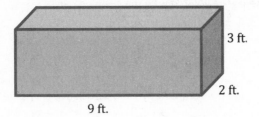

3 ft.

2 ft.

9 ft.

a. Determine the volume of the rectangular prism.

b. Determine the volume if all of the sides are half as long as the original lengths.

c. Determine the volume if all of the sides are one-third as long as the original lengths.

d. Is the relationship between the side lengths and the volume the same as the one that occurred in Exercise 3? Explain your answer.

Lesson 13: The Formulas for Volume

EUREKA MATH

5.

    a.   If e represents a side length of the cube, create an expression that shows the volume of the cube.

    b.   If we divide the side lengths by three, create an expression for the new side length.

    c.   Write an expression that represents the volume of the cube with one-third the side length.

    d.   Write an equivalent expression to show that the volume is $\frac{1}{27}$ of the original volume.

Name _____   Date _____

1. A new company wants to mail out samples of its hair products. The company has a sample box that is a rectangular prism with a rectangular base with an area of $23\frac{1}{3}$ in². The height of the prism is $1\frac{1}{4}$ in.

Determine the volume of the sample box.

2. A different sample box has a height that is twice as long as the original box described in Problem 1. What is the volume of this sample box? How does the volume of this sample box compare to the volume of the sample box in Problem 1?

1.

a. Determine the volume of a cube with a side length of $3\frac{3}{4}$ cm.

$$V = l\,w\,h$$

$$V = \left(3\frac{3}{4}\text{ cm}\right)\left(3\frac{3}{4}\text{ cm}\right)\left(3\frac{3}{4}\text{ cm}\right)$$

> I know that if the prism is a cube, the length, width, and height all have the same measure.

$$V = \left(\frac{15}{4}\text{ cm}\right)\left(\frac{15}{4}\text{ cm}\right)\left(\frac{15}{4}\text{ cm}\right)$$

$$V = \frac{3,375}{64}\text{ cm}^3$$

The volume of the cube is $\frac{3,375}{64}$ cm^3.

b. Determine the volume of the cube in part (a) if all of the side lengths are doubled.

$$3\frac{3}{4}\text{ cm} \times 2 = 7\frac{1}{2}\text{ cm}$$

> I must multiply the original dimensions by 2 to determine the dimensions of the new cube.

$$V = \left(7\frac{1}{2}\text{ cm}\right)\left(7\frac{1}{2}\text{ cm}\right)\left(7\frac{1}{2}\text{ cm}\right)$$

$$V = \left(\frac{15}{2}\text{ cm}\right)\left(\frac{15}{2}\text{ cm}\right)\left(\frac{15}{2}\text{ cm}\right)$$

$$V = \frac{3,375}{8}\text{ cm}^3$$

The volume of this cube is $\frac{3,375}{8}$ cm^3.

c. Determine the relationship between the volumes in part (a) and part (b).

The volume of the prism in part (b) is 8 times greater than the volume of the prism in part (a).

> I know the length is twice as long, the width is twice as long, and the height is twice as long as the original cube.
> $V = (2l)(2w)(2h) = 8\,l\,w\,h$

2. Use the rectangular prism to answer the following questions.

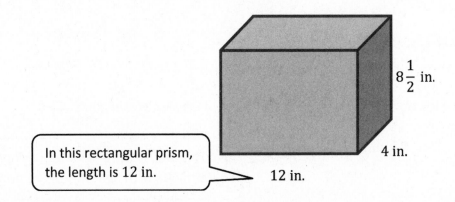

> In this rectangular prism, the length is 12 in.

$8\frac{1}{2}$ in.

4 in.

12 in.

a. Complete the table.

Length	Volume
$l = 12$ in.	**408 in^3**
$\frac{1}{2}l = 6$ in.	**204 in^3**
$\frac{1}{3}l = 4$ in.	**136 in^3**
$\frac{1}{4}l = 3$ in.	**102 in^3**
$2l = 24$ in.	**816 in^3**
$3l = 36$ in.	**1,224 in^3**
$4l = 48$ in.	**1,632 in^3**

> When I calculate the volumes in the table, the length changes, but the width and height stay the same.

EUREKA MATH

b. How did the volume change when the length was one-fourth as long as the original length?

When the length is one-fourth as long as the original length, the volume is one-fourth as much as the original volume.

I need to compare the new volume to the original volume. I can find the quotient of 102 in^3 and 408 in^3 to determine the change in volumes.

c. How did the volume change when the length was four times as long as the original length?

When the length is four times as long as the original length, the volume is four times as much as the original volume.

I know that the difference in the volumes can be shown using subtraction.

3. The difference between the volumes of two rectangular prisms, Box A and Box B, is 23.87 cm^3. Box B has a volume of 34.69 cm^3.

a. Let A represent the volume of Box A in cubic centimeters. Write an equation that could be used to determine the volume of Box A.

$$A - 34.69 \text{ cm}^3 = 23.87 \text{ cm}^3$$

b. Solve the equation to determine the volume of Box A.

$$A - 34.69 \text{ cm}^3 = 23.87 \text{ cm}^3$$
$$A - 34.69 \text{ cm}^3 + 34.69 \text{ cm}^3 = 23.87 \text{ cm}^3 + 34.69 \text{ cm}^3$$
$$A = 58.56 \text{ cm}^3$$

The volume of Box A is 58.56 cm^3.

c. If the area of the base of Box A is 4 cm^2, write an equation that could be used to determine the height of Box A. Let h represent the height of Box A in centimeters.

$$V = Bh$$
$$58.56 \text{ cm}^3 = (4 \text{ cm}^2)h$$

d. Solve the equation to determine the height of Box A.

$$58.56 \text{ cm}^3 = (4 \text{ cm}^2)h$$

$$58.56 \text{ cm}^3 \div 4 \text{ cm}^2 = (4 \text{ cm}^2)h \div 4 \text{ cm}^2$$

$$14.64 \text{ cm} = h$$

The height of Box A is 14.64 cm.

Lesson 13:   The Formulas for Volume

EUREKA
MATH®

1. Determine the volume of the rectangular prism.

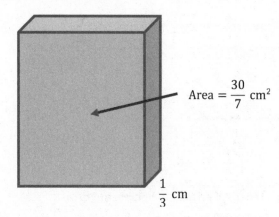

$$\text{Area} = \frac{30}{7} \text{ cm}^2$$

$\frac{1}{3}$ cm

2. Determine the volume of the rectangular prism in Problem 1 if the height is quadrupled (multiplied by four). Then, determine the relationship between the volumes in Problem 1 and this prism.

3. The area of the base of a rectangular prism can be represented by B, and the height is represented by h.

   a. Write an equation that represents the volume of the prism.

   b. If the area of the base is doubled, write an equation that represents the volume of the prism.

   c. If the height of the prism is doubled, write an equation that represents the volume of the prism.

   d. Compare the volume in parts (b) and (c). What do you notice about the volumes?

   e. Write an expression for the volume of the prism if both the height and the area of the base are doubled.

4. Determine the volume of a cube with a side length of $5\frac{1}{3}$ in.

5. Use the information in Problem 4 to answer the following:

   a. Determine the volume of the cube in Problem 4 if all of the side lengths are cut in half.

   b. How could you determine the volume of the cube with the side lengths cut in half using the volume in Problem 4?

6. Use the rectangular prism to answer the following questions.

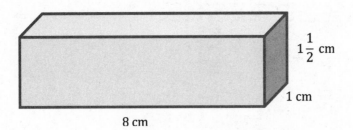

$1\frac{1}{2}$ cm

1 cm

8 cm

a. Complete the table.

Length of Prism	Volume of Prism
$l = 8$ cm	
$\frac{1}{2}l =$	
$\frac{1}{3}l =$	
$\frac{1}{4}l =$	
$2l =$	
$3l =$	
$4l =$	

b. How did the volume change when the length was one-third as long?

c. How did the volume change when the length was tripled?

d. What conclusion can you make about the relationship between the volume and the length?

7. The sum of the volumes of two rectangular prisms, Box A and Box B, are 14.325 cm^3. Box A has a volume of 5.61 cm^3.

a. Let B represent the volume of Box B in cubic centimeters. Write an equation that could be used to determine the volume of Box B.

b. Solve the equation to determine the volume of Box B.

c. If the area of the base of Box B is 1.5 cm^2, write an equation that could be used to determine the height of Box B. Let h represent the height of Box B in centimeters.

d. Solve the equation to determine the height of Box B.

EUREKA MATH®

Example 1

a. The area of the base of a sandbox is $9\frac{1}{2}$ ft^2. The volume of the sandbox is $7\frac{1}{8}$ ft^3. Determine the height of the sandbox.

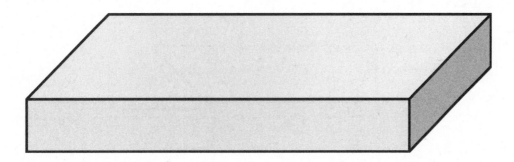

b. The sandbox was filled with sand, but after the kids played, some of the sand spilled out. Now, the sand is at a height of $\frac{1}{2}$ ft. Determine the volume of the sand in the sandbox after the children played in it.

Example 2

A special-order sandbox has been created for children to use as an archeological digging area at the zoo. Determine the volume of the sandbox.

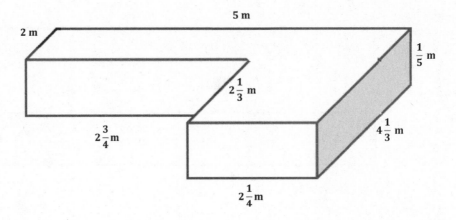

Exercises

1.

a. The volume of the rectangular prism is $\frac{35}{15}$ yd^3. Determine the missing measurement using a one-step equation.

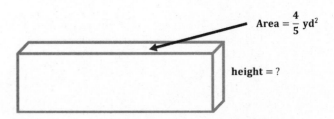

Area $= \frac{4}{5}$ yd^2

height $= ?$

EUREKA
MATH

b.   The volume of the box is $\frac{45}{6}$ m³. Determine the area of the base using a one-step equation.

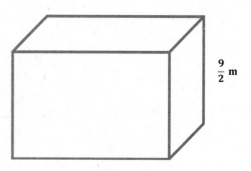

$\frac{9}{2}$ m

2.   Marissa's fish tank needs to be filled with more water.

a.   Determine how much water the tank can hold.

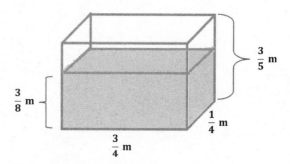

$\frac{3}{5}$ m

$\frac{3}{8}$ m

$\frac{1}{4}$ m

$\frac{3}{4}$ m

b.   Determine how much water is already in the tank.

c.   How much more water is needed to fill the tank?

3. Determine the volume of the composite figures.

a.

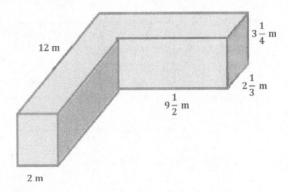

b.

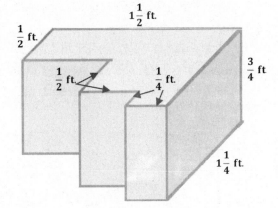

Lesson 14:   Volume in the Real World

EUREKA
MATH·

Name _____   Date _____

1. Determine the volume of the water that would be needed to fill the rest of the tank.

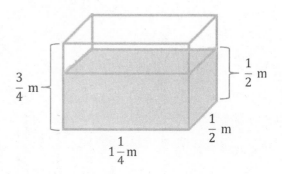

2. Determine the volume of the composite figure.

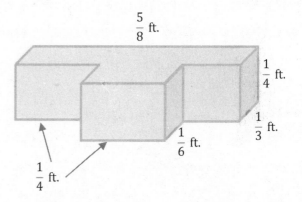

1. The volume of a rectangular prism is $\frac{35}{24}$ cm^3, and the height of the prism is $\frac{7}{4}$ cm. Determine the area of the base.

Area of base = volume ÷ height

Area of base $= \dfrac{35}{24}$ **cm^3** $\div \dfrac{7}{4}$ **cm**

Area of base $= \dfrac{35}{24}$ **cm^3** $\div \dfrac{42}{24}$ **cm**

Area of base $= 35$ **cm^3** $\div 42$ **cm**

Area of base $= \dfrac{35}{42}$ **cm^2**

Area of base $= \dfrac{5}{6}$ **cm^2**

The area of the base is $\dfrac{5}{6}$ *cm^2.*

> I can use the equation Volume = Area of base × height and then solve for the area of the base to get a formula to use in this question.

> I remember from Module 2 that I can divide the numerators when I have common denominators.

2. Determine the volume of the space in the tank that still needs to be filled with water.

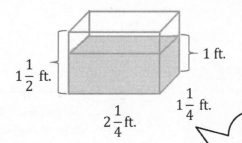

$1\frac{1}{2}$ ft.

1 ft.

$1\frac{1}{4}$ ft.

$2\frac{1}{4}$ ft.

> I can determine the volume of the tank and the volume of the water that is already in the tank. The difference between the two will be the amount needed to fill the rest of the tank.

Volume of tank $= l\ w\ h$

Volume of tank $= \left(2\dfrac{1}{4} \text{ ft.}\right)\left(1\dfrac{1}{4} \text{ ft.}\right)\left(1\dfrac{1}{2} \text{ ft.}\right)$

Volume of tank $= \left(\dfrac{9}{4} \text{ ft.}\right)\left(\dfrac{5}{4} \text{ ft.}\right)\left(\dfrac{3}{2} \text{ ft.}\right)$

Volume of tank $= \dfrac{135}{32}$ **ft^3**

Volume of water = $l\ w\ h$

Volume of water = $\left(2\dfrac{1}{4}\text{ ft.}\right)\left(1\dfrac{1}{4}\text{ ft.}\right)(1\text{ ft.})$

Volume of water = $\left(\dfrac{9}{4}\text{ ft.}\right)\left(\dfrac{5}{4}\text{ ft.}\right)(1\text{ ft.})$

Volume of water = $\dfrac{45}{16}\text{ ft}^3$

Remaining water needed:

$\dfrac{135}{32}\text{ ft}^3 - \dfrac{45}{16}\text{ ft}^3 = \dfrac{135}{32}\text{ ft}^3 - \dfrac{90}{32}\text{ ft}^3 = \dfrac{45}{32}\text{ ft}^3$

The volume of the space that still needs to be filled is $\dfrac{45}{32}$ ft³.

3.

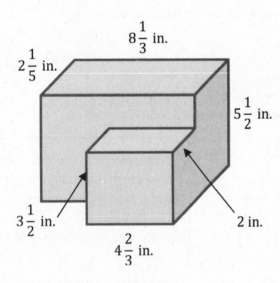

This looks like two rectangular prisms that were stuck together.

I can find the volume of each piece and then add the volumes together.

a. Write an equation to represent the volume of the composite figure.

$$V = \left(8\dfrac{1}{3}\text{ in.} \times 2\dfrac{1}{5}\text{ in.} \times 5\dfrac{1}{2}\text{ in.}\right) + \left(4\dfrac{2}{3}\text{ in.} \times 2\text{ in.} \times 3\dfrac{1}{2}\text{ in.}\right)$$

EUREKA MATH®

b. Use your equation to calculate the volume of the composite figure.

$$V = \left(8\frac{1}{3} \text{ in.} \times 2\frac{1}{5} \text{ in.} \times 5\frac{1}{2} \text{ in.}\right) + \left(4\frac{2}{3} \text{ in.} \times 2 \text{ in.} \times 3\frac{1}{2} \text{ in.}\right)$$

$$V = \left(\frac{25}{3} \text{ in.} \times \frac{11}{5} \text{ in.} \times \frac{11}{2} \text{ in.}\right) + \left(\frac{14}{3} \text{ in.} \times \frac{2}{1} \text{ in.} \times \frac{7}{2} \text{ in.}\right)$$

$$V = \frac{605}{6} \text{ in}^3 + \frac{196}{6} \text{ in}^3$$

$$V = \frac{801}{6} \text{ in}^3$$

$$V = 133\frac{1}{2} \text{ in}^3$$

The volume of the composite figure is $133\frac{1}{2}$ *in*^{3}.

1. The volume of a rectangular prism is $\frac{21}{12}$ ft³, and the height of the prism is $\frac{3}{4}$ ft. Determine the area of the base.

2. The volume of a rectangular prism is $\frac{10}{21}$ ft³. The area of the base is $\frac{2}{3}$ ft². Determine the height of the rectangular prism.

3. Determine the volume of the space in the tank that still needs to be filled with water if the water is $\frac{1}{3}$ ft. deep.

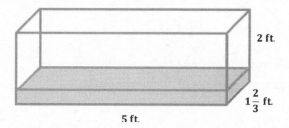

4. Determine the volume of the composite figure.

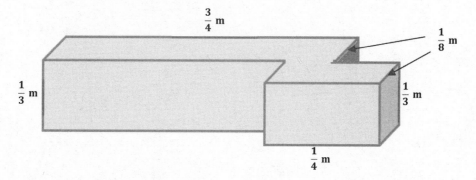

5. Determine the volume of the composite figure.

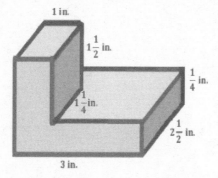

6.

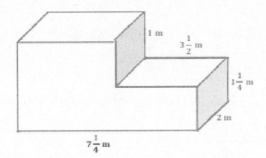

a.   Write an equation to represent the volume of the composite figure.

b.   Use your equation to calculate the volume of the composite figure.

Lesson 14:   Volume in the Real World

EUREKA
MATH

Exercise: Cube

1. Nets are two-dimensional figures that can be folded into three-dimensional solids. Some of the drawings below are nets of a cube. Others are not cube nets; they can be folded, but not into a cube.

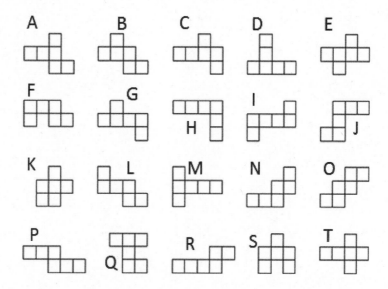

a. Experiment with the larger cut-out patterns provided. Shade in each of the figures above that can fold into a cube.

b. Write the letters of the figures that can be folded into a cube.

c. Write the letters of the figures that cannot be folded into a cube.

Lesson Summary

NET: If the surface of a 3-dimensional solid can be cut along sufficiently many edges so that the faces can be placed in one plane to form a connected figure, then the resulting system of faces is called a *net of the solid*.

Name _____   Date _____

1.   What is a net? Describe it in your own words.

2.   Which of the following can fold to make a cube? Explain how you know.

1. Match the net below to the picture of its solid.

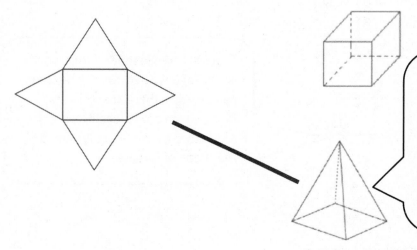

I know this shape is a pyramid because there are multiple identical triangles that form the lateral (side) faces of the pyramid, while the remaining face is the base (rectangle).

2. Sketch a net that can fold into a cube.

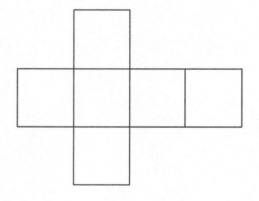

I know there are many possible nets for a cube. I can draw this net because I know a cube has six square faces. I can strategically draw the faces so the net, when folded along the edges, will create a closed, solid figure (in this case a cube). The image below shows what the net would look like when it's folded to create a cube.

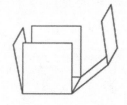

3. Below is the net for a prism or pyramid. Classify the solid as prism or pyramid, and identify the shape of the base(s). Then, write the name of the solid.

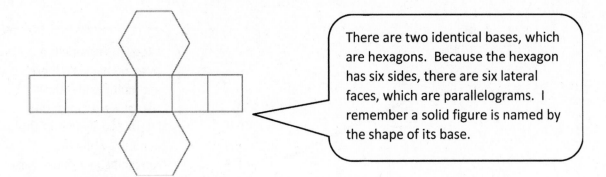

> There are two identical bases, which are hexagons. Because the hexagon has six sides, there are six lateral faces, which are parallelograms. I remember a solid figure is named by the shape of its base.

This solid is a prism, and the bases are hexagons.

This is a hexagonal prism.

EUREKA
MATH

1. Match the following nets to the picture of its solid. Then, write the name of the solid.

a.

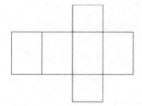

d.

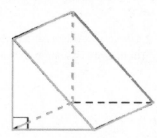

b.

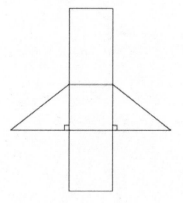

e.

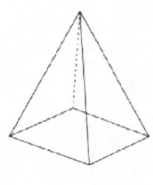

c.

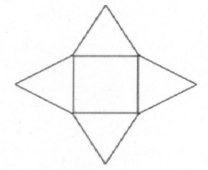

f.

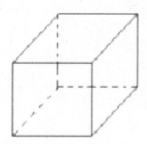

2. Sketch a net that can fold into a cube.

3. Below are the nets for a variety of prisms and pyramids. Classify the solids as prisms or pyramids, and identify the shape of the base(s). Then, write the name of the solid.

a.

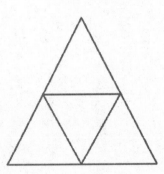

b.

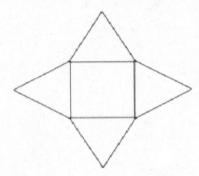

c.

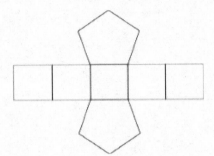

d.

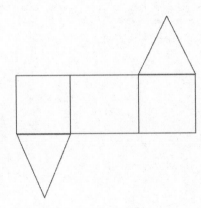

e.

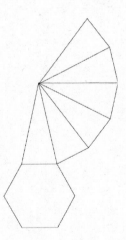

f.

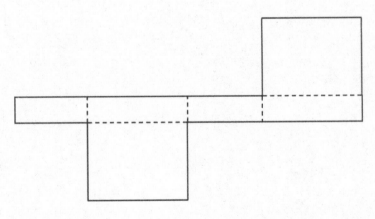

EUREKA
MATH®

Opening Exercise

Sketch the faces in the area below. Label the dimensions.

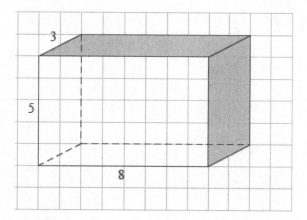

Exploratory Challenge 1: Rectangular Prisms

a. Use the measurements from the solid figures to cut and arrange the faces into a net. (Note: All measurements are in centimeters.)

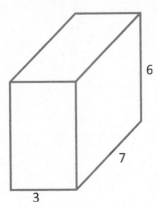

b. A juice box measures 4 inches high, 3 inches long, and 2 inches wide. Cut and arrange all 6 faces into a net. (Note: All measurements are in inches.)

c. Challenge: Write a numerical expression for the total area of the net for part (b). Explain each term in your expression.

EUREKA
MATH®

Exploratory Challenge 2: Triangular Prisms

Use the measurements from the triangular prism to cut and arrange the faces into a net. (Note: All measurements are in inches.)

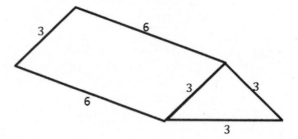

Exploratory Challenge 3: Pyramids

Pyramids are named for the shape of the base.

a. Use the measurements from this square pyramid to cut and arrange the faces into a net. Test your net to be sure it folds into a square pyramid.

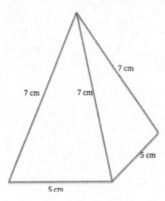

b. A triangular pyramid that has equilateral triangles for faces is called a tetrahedron. Use the measurements from this tetrahedron to cut and arrange the faces into a net.

All edges are 4 in. in length.

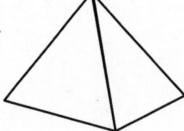

EUREKA
MATH

Name _____   Date _____

Sketch and label a net of this pizza box. It has a square top that measures 16 inches on a side, and the height is 2 inches. Treat the box as a prism, without counting the interior flaps that a pizza box usually has.

1. Sketch and label the net of the following solid figures, and label the edge lengths.

   a.   A granola cereal box that measures 11 inches high, 5 inches long, and 2 inches wide

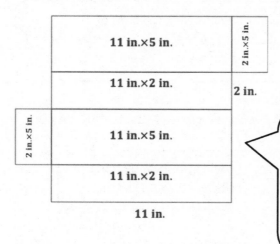

I can use the dimensions of this rectangular prism (granola cereal box) to create the net. There are three pairs of faces with identical areas, and I can arrange them so that this net will fold into a rectangular prism. The dimensions help me label the edge lengths.

   b.   A cubic box that measures 12 cm on each edge.

   All edges are 12 cm.

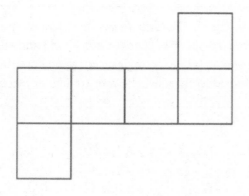

The net of a cube has 6 identical square faces. This is one possible arrangement of the faces.

c. Challenge: Write a numerical expression for the total area of the net in part (b). Tell what each of the terms in your expression means.

6(12 cm × 12 cm)

> This is how I calculate the area of one face.

> This is another way of saying 6 groups of (12 cm × 12 cm) since there are 6 congruent faces and each face is a square with an edge of 12 cm.

(12 cm × 12 cm) + (12 cm × 12 cm) + (12 cm × 12 cm) + (12 cm × 12 cm) + (12 cm × 12 cm) + (12 cm × 12 cm)

> I could also show 6 groups of (12 cm × 12 cm) like this.

There are 6 faces in the cube, and each has dimensions 12 cm by 12 cm.

2. The base for a patio umbrella is shaped like a square pyramid. The pyramid has equilateral faces that measure 17 inches on each side. The base is 17 inches on each side. Sketch the net of the umbrella base, and label the edge lengths.

Possible Net:

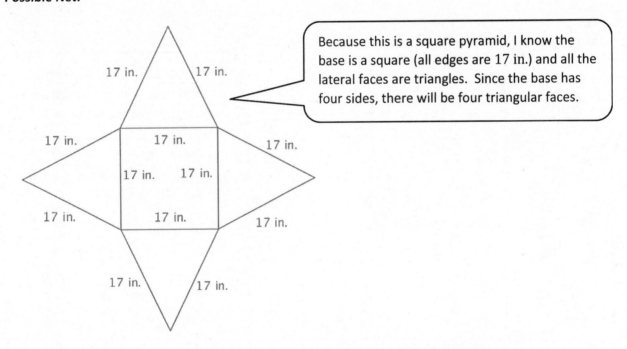

> Because this is a square pyramid, I know the base is a square (all edges are 17 in.) and all the lateral faces are triangles. Since the base has four sides, there will be four triangular faces.

EUREKA MATH®

3. The roof of a detached garage is in the shape of a triangular prism. It has equilateral bases that measure 9 feet on each side. The length of the roof is 23 feet. Sketch the net of the detached garage, and label the edge lengths.

    Possible Net:

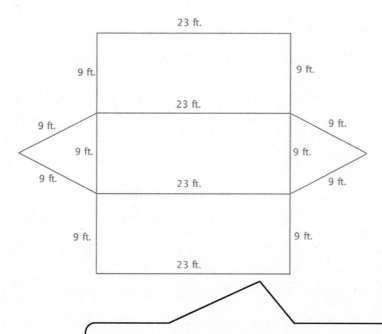

Since this is a triangular prism, I know there are two triangular bases. Since the base is a triangle and has three sides, there will be three rectangular faces.

1. Sketch and label the net of the following solid figures, and label the edge lengths.

   a. A cereal box that measures 13 inches high, 7 inches long, and 2 inches wide

   b. A cubic gift box that measures 8 cm on each edge

   c. Challenge: Write a numerical expression for the total area of the net in part (b). Tell what each of the terms in your expression means.

2. This tent is shaped like a triangular prism. It has equilateral bases that measure 5 feet on each side. The tent is 8 feet long. Sketch the net of the tent, and label the edge lengths.

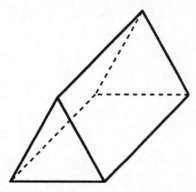

3. The base of a table is shaped like a square pyramid. The pyramid has equilateral faces that measure 25 inches on each side. The base is 25 inches long. Sketch the net of the table base, and label the edge lengths.

4. The roof of a shed is in the shape of a triangular prism. It has equilateral bases that measure 3 feet on each side. The length of the roof is 10 feet. Sketch the net of the roof, and label the edge lengths.

Number Correct: _____

Addition and Subtraction Equations—Round 1

Directions: Find the value of m in each equation.

1.	$m + 4 = 11$	
2.	$m + 2 = 5$	
3.	$m + 5 = 8$	
4.	$m - 7 = 10$	
5.	$m - 8 = 1$	
6.	$m - 4 = 2$	
7.	$m + 12 = 34$	
8.	$m + 25 = 45$	
9.	$m + 43 = 89$	
10.	$m - 20 = 31$	
11.	$m - 13 = 34$	
12.	$m - 45 = 68$	
13.	$m + 34 = 41$	
14.	$m + 29 = 52$	
15.	$m + 37 = 61$	
16.	$m - 43 = 63$	
17.	$m - 21 = 40$	

18.	$m - 54 = 37$	
19.	$4 + m = 9$	
20.	$6 + m = 13$	
21.	$2 + m = 31$	
22.	$15 = m + 11$	
23.	$24 = m + 13$	
24.	$32 = m + 28$	
25.	$4 = m - 7$	
26.	$3 = m - 5$	
27.	$12 = m - 14$	
28.	$23.6 = m - 7.1$	
29.	$14.2 = m - 33.8$	
30.	$2.5 = m - 41.8$	
31.	$64.9 = m + 23.4$	
32.	$72.2 = m + 38.7$	
33.	$1.81 = m - 15.13$	
34.	$24.68 = m - 56.82$	

EUREKA MATH®

Number Correct: _____

Improvement: _____

Addition and Subtraction Equations—Round 2

Directions: Find the value of m in each equation.

1.	$m + 2 = 7$	
2.	$m + 4 = 10$	
3.	$m + 8 = 15$	
4.	$m + 7 = 23$	
5.	$m + 12 = 16$	
6.	$m - 5 = 2$	
7.	$m - 3 = 8$	
8.	$m - 4 = 12$	
9.	$m - 14 = 45$	
10.	$m + 23 = 40$	
11.	$m + 13 = 31$	
12.	$m + 23 = 48$	
13.	$m + 38 = 52$	
14.	$m - 14 = 27$	
15.	$m - 23 = 35$	
16.	$m - 17 = 18$	
17.	$m - 64 = 1$	

18.	$6 = m + 3$	
19.	$12 = m + 7$	
20.	$24 = m + 16$	
21.	$13 = m + 9$	
22.	$32 = m - 3$	
23.	$22 = m - 12$	
24.	$34 = m - 10$	
25.	$48 = m + 29$	
26.	$21 = m + 17$	
27.	$52 = m + 37$	
28.	$\dfrac{6}{7} = m + \dfrac{4}{7}$	
29.	$\dfrac{2}{3} = m - \dfrac{5}{3}$	
30.	$\dfrac{1}{4} = m - \dfrac{8}{3}$	
31.	$\dfrac{5}{6} = m - \dfrac{7}{12}$	
32.	$\dfrac{7}{8} = m - \dfrac{5}{12}$	
33.	$\dfrac{7}{6} + m = \dfrac{16}{3}$	
34.	$\dfrac{1}{3} + m = \dfrac{13}{15}$	

Opening Exercise

    a.   Write a numerical equation for the area of the figure below. Explain and identify different parts of the figure.

        i.

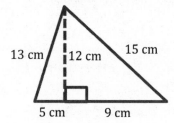

        ii.   How would you write an equation that shows the area of a triangle with base b and height h?

    b.   Write a numerical equation for the area of the figure below. Explain and identify different parts of the figure.

        i.

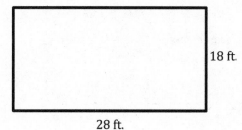

        ii.   How would you write an equation that shows the area of a rectangle with base b and height h?

Example 1

Use the net to calculate the surface area of the figure. (Note: all measurements are in centimeters.)

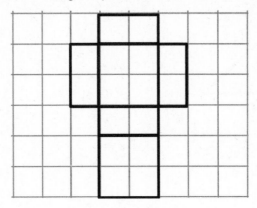

Example 2

Use the net to write an expression for surface area. (Note: all measurements are in square feet.)

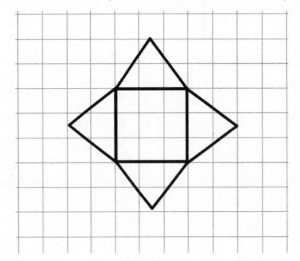

Lesson 17:   From Nets to Surface Area

EUREKA
MATH®

Exercises

Name the solid the net would create, and then write an expression for the surface area. Use the expression to determine the surface area. Assume that each box on the grid paper represents a 1 cm × 1 cm square. Explain how the expression represents the figure.

1.

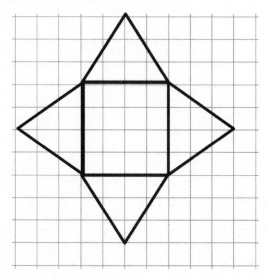

2.

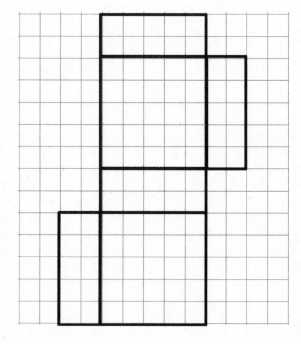

3.

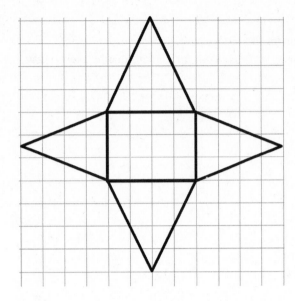

4.

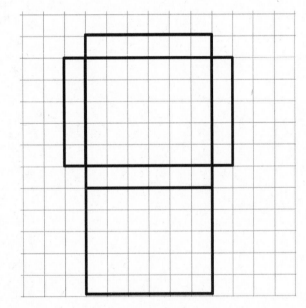

Lesson 17:   From Nets to Surface Area

EUREKA
MATH

Name _____   Date _____

Name the shape, and then calculate the surface area of the figure. Assume each box on the grid paper represents a 1 in. × 1 in. square.

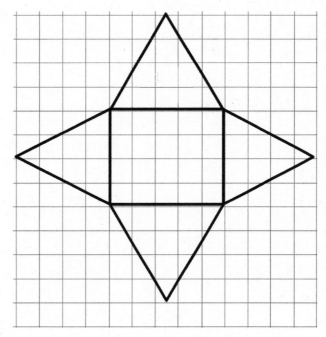

Name the shape, and write an expression for surface area. Calculate the surface area of the figure. Assume each box on the grid paper represents a 1 ft. × 1 ft. square.

1.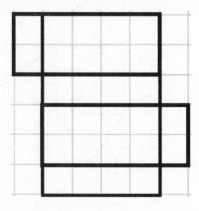

Name of Shape: Rectangular Prism

> I know this is a rectangular prism because there are six rectangular faces.

> To find the surface area, I can find the area of each face and then find the sum of the areas of all six faces.

Surface Area:

$(2 \text{ ft.} \times 1 \text{ ft.}) + (2 \text{ ft.} \times 1 \text{ ft.}) + (4 \text{ ft.} \times 2 \text{ ft.}) + (4 \text{ ft.} \times 2 \text{ ft.}) + (4 \text{ ft.} \times 1 \text{ ft.}) + (4 \text{ ft.} \times 1 \text{ ft.})$

$= 2(2 \text{ ft.} \times 1 \text{ ft.}) + 2(4 \text{ ft.} \times 2 \text{ ft.}) + 2(4 \text{ ft.} \times 1 \text{ ft.})$

$= 4 \text{ ft}^2 + 16 \text{ ft}^2 + 8 \text{ ft}^2$

$= 28 \text{ ft}^2$

> I notice there are three groups of faces that have identical areas, so I can rewrite the expression to reflect this idea.

Explain the error in Problems 2 and 3. Then, correct the error. Assume each box on the grid paper represents a 1 cm × 1 cm square.

2. Name of Shape: Rectangular Prism or, more specifically, a Cube

Area of Faces: $2 \text{ cm} \times 2 \text{ cm} = 4 \text{ cm}^2$

Surface Area: $4 \text{ cm}^2 + 4 \text{ cm}^2 + 4 \text{ cm}^2 + 4 \text{ cm}^2 = 16 \text{ cm}^2$

The solution shown above only calculates the sum of four faces, but a cube has six faces so the solution above is incorrect.

$4 \text{ cm}^2 + 4 \text{ cm}^2 + 4 \text{ cm}^2 + 4 \text{ cm}^2 + 4 \text{ cm}^2 + 4 \text{ cm}^2 = 24 \text{ cm}^2$

Therefore, the correct surface area is 24 cm^2 and not 16 cm^2.

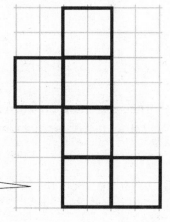

> I see there are six faces when looking at the net, but the work for surface area only shows the sum of four faces.

3. Name of Shape: Rectangular Pyramid, but more specifically, a Square Pyramid

    Area of Base: $4\,\text{m} \times 4\,\text{m} = 12\,\text{m}^2$

    Area of Triangles: $\frac{1}{2} \times 4\,\text{m} \times 3\,\text{m} = 6\,\text{m}^2$

    Surface Area: $12\,\text{m}^2 + 6\,\text{m}^2 + 6\,\text{m}^2 + 6\,\text{m}^2 + 6\,\text{m}^2 = 36\,\text{m}^2$

    The method used to calculate the surface area is correct, but there is a math error in the calculation of the area of the base. The area of the base should be $16\,\text{m}^2$ because the product of $4\,\text{m}$ and $4\,\text{m}$ is $16\,\text{m}^2$.

    $$16\,\text{m}^2 + 6\,\text{m}^2 + 6\,\text{m}^2 + 6\,\text{m}^2 + 6\,\text{m}^2 = 40\,\text{m}^2$$

    Therefore, the correct surface area of the square pyramid is $40\,\text{m}^2$.

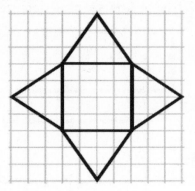

4. Catherine and Julia are both writing expressions to calculate the surface area of a rectangular prism. However, they wrote different expressions.

    a. Examine the expressions below, and determine if they represent the same value. Explain why or why not.

    Catherine's Expression

    $$(6\,\text{in.} \times 9\,\text{in.}) + (6\,\text{in.} \times 9\,\text{in.}) + (6\,\text{in.} \times 10\,\text{in.}) + (6\,\text{in.} \times 10\,\text{in.}) + (9\,\text{in.} \times 10\,\text{in.}) + (9\,\text{in.} \times 10\,\text{in.})$$

    Julia's Expression

    $$2(6\,\text{in.} \times 9\,\text{in.}) + 2(6\,\text{in.} \times 10\,\text{in.}) + 2(9\,\text{in.} \times 10\,\text{in.})$$

    > Julia realized there are three groups of identical products, so she combined these using the distributive property.

    The expressions are equivalent expressions because they have the same value, but Julia used the distributive property to write a more compact expression than Catherine's.

    b. What fact about the surface area of a rectangular prism does Julia's expression show that Catherine's does not?

    A rectangular prism is composed of three pairs of sides with identical areas.

Lesson 17:   From Nets to Surface Area

EUREKA MATH

Name the shape, and write an expression for surface area. Calculate the surface area of the figure. Assume each box on the grid paper represents a 1 ft. × 1 ft. square.

1.

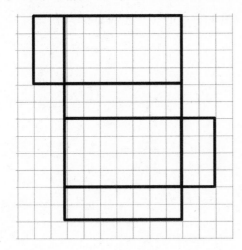

2.

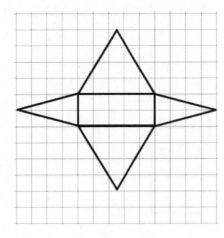

Explain the error in each problem below. Assume each box on the grid paper represents a 1 m × 1 m square.

3.

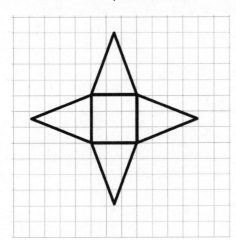

Name of Shape: Rectangular Pyramid, but more specifically a Square Pyramid

Area of Base: $3\,\text{m} \times 3\,\text{m} = 9\,\text{m}^2$

Area of Triangles: $3\,\text{m} \times 4\,\text{m} = 12\,\text{m}^2$

Surface Area: $9\,\text{m}^2 + 12\,\text{m}^2 + 12\,\text{m}^2 + 12\,\text{m}^2 + 12\,\text{m}^2 = 57\,\text{m}^2$

4.

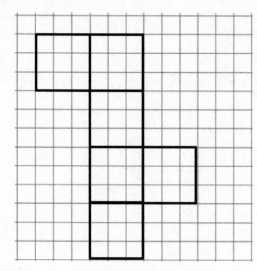

Name of Shape: Rectangular Prism or, more specifically, a Cube

Area of Faces: $3 \text{ m} \times 3 \text{ m} = 9 \text{ m}^2$

Surface Area: $9 \text{ m}^2 + 9 \text{ m}^2 + 9 \text{ m}^2 + 9 \text{ m}^2 + 9 \text{ m}^2 = 45 \text{ m}^2$

5. Sofia and Ella are both writing expressions to calculate the surface area of a rectangular prism. However, they wrote different expressions.

   a. Examine the expressions below, and determine if they represent the same value. Explain why or why not.

Sofia's Expression:

$$(3 \text{ cm} \times 4 \text{ cm}) + (3 \text{ cm} \times 4 \text{ cm}) + (3 \text{ cm} \times 5 \text{ cm}) + (3 \text{ cm} \times 5 \text{ cm}) + (4 \text{ cm} \times 5 \text{ cm}) + (4 \text{ cm} \times 5 \text{ cm})$$

Ella's Expression:

$$2(3 \text{ cm} \times 4 \text{ cm}) + 2(3 \text{ cm} \times 5 \text{ cm}) + 2(4 \text{ cm} \times 5 \text{ cm})$$

   b. What fact about the surface area of a rectangular prism does Ella's expression show more clearly than Sofia's?

Lesson 17:   From Nets to Surface Area

EUREKA MATH

Opening Exercise

  a. What three-dimensional figure does the net create?

  b. Measure (in inches) and label each side of the figure.

  c. Calculate the area of each face, and record this value inside the corresponding rectangle.

  d. How did we compute the surface area of solid figures in previous lessons?

  e. Write an expression to show how we can calculate the surface area of the figure above.

  f. What does each part of the expression represent?

  g. What is the surface area of the figure?

Example 1

Fold the net used in the Opening Exercise to make a rectangular prism. Have the two faces with the largest area be the bases of the prism. Fill in the first row of the table below.

Area of Top (base)	Area of Bottom (base)	Area of Front	Area of Back	Area of Left Side	Area of Right Side

Examine the rectangular prism below. Complete the table.

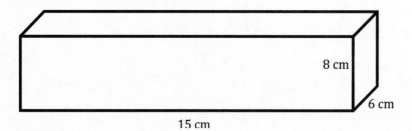

Area of Top (base)	Area of Bottom (base)	Area of Front	Area of Back	Area of Left Side	Area of Right Side

Lesson 18:   Determining Surface Area of Three-Dimensional
                         Figures

EUREKA MATH

Example 2

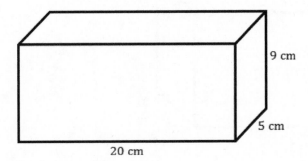

9 cm

5 cm

20 cm

Exercises 1–3

1. Calculate the surface area of each of the rectangular prisms below.

a.

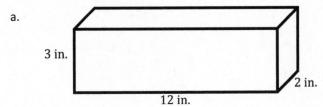

3 in.

12 in.

2 in.

b.

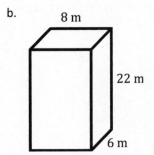

8 m

22 m

6 m

c.

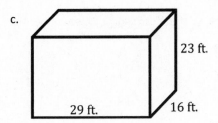

23 ft.

29 ft.

16 ft.

EUREKA MATH®

d.

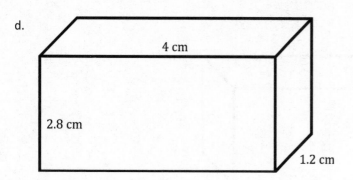

2. Calculate the surface area of the cube.

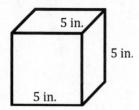

3. All the edges of a cube have the same length. Tony claims that the formula $SA = 6s^2$, where s is the length of each side of the cube, can be used to calculate the surface area of a cube.

    a. Use the dimensions from the cube in Problem 2 to determine if Tony's formula is correct.

    b. Why does this formula work for cubes?

    c. Becca does not want to try to remember two formulas for surface area, so she is only going to remember the formula for a cube. Is this a good idea? Why or why not?

EUREKA
MATH

Lesson Summary

Surface Area Formula for a Rectangular Prism: $SA = 2lw + 2lh + 2wh$

Surface Area Formula for a Cube: $SA = 6s^2$

Name _____   Date _____

Calculate the surface area of each figure below. Figures are not drawn to scale.

1.

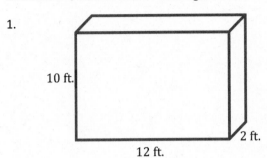

10 ft.

12 ft.

2 ft.

2.

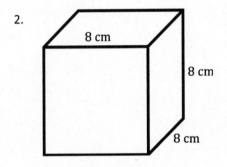

8 cm

8 cm

8 cm

Calculate the surface area of the figures below. This figures are not drawn to scale.

1.

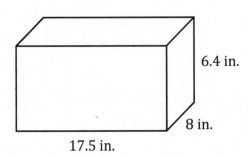

6.4 in.

8 in.

17.5 in.

The formula $SA = 2(l \times w) + 2(l \times h) + 2(w \times h)$ can be used to determine the surface area of this figure.

$$SA = 2(17.5 \text{ in.})(8 \text{ in.}) + 2(17.5 \text{ in.})(6.4 \text{ in.}) + 2(8 \text{ in.})(6.4 \text{ in.})$$

$$SA = 280 \text{ in}^2 + 224 \text{ in}^2 + 102.4 \text{ in}^2$$

$$SA = 606.4 \text{ in}^2$$

The surface area of the figure is 606.4 in².

2.

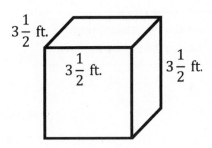

$3\frac{1}{2}$ ft.

$3\frac{1}{2}$ ft.

$3\frac{1}{2}$ ft.

All six faces of a cube are identical squares, which means I can square the side length since the side lengths are the same and then multiply by 6.

$$SA = 6\left(3\frac{1}{2} \text{ ft.}\right)^2$$

$$SA = 6\left(\frac{7}{2} \text{ ft.}\right)^2$$

$$SA = 6\left(\frac{49}{4} \text{ ft}^2\right)$$

$$SA = \frac{294}{4} \text{ ft}^2$$

$$SA = 73\frac{1}{2} \text{ ft}^2$$

The surface area of the figure is $73\frac{1}{2}$ ft².

3. Write a numerical expression to show how to calculate the surface area of the rectangular prism. Explain each part of the expression.

    $2(4 \text{ ft.} \times 11 \text{ ft.}) + 2(4 \text{ ft.} \times 2 \text{ ft.}) + 2(11 \text{ ft.} \times 2 \text{ ft.})$

    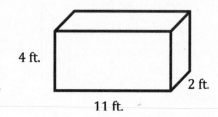

    The first term of the expression shows the area of the front and back faces of the rectangular prism. The second term of the expression shows the area of the left and right faces of the rectangular prism. The third term shows the area of the top and bottom faces of the rectangular prism.

    The surface area of the figure is 148 ft^2*.*

4. When Annabelle was calculating the surface area for Problem 1, she identified the following:

    length = 17.5 in., width = 8 in., and height = 6.4 in.

    However, when Vincent was calculating the surface area for the same problem, he identified the following:

    length = 8 in., width = 17.5 in., and height = 6.4 in.

    Would Annabelle and Vincent get the same answer? Why or why not?

    Annabelle and Vincent would get the same answer because they are still finding the correct area of all six faces of the rectangular prism.

    > The surface area will be the same because the dimensions represent the same rectangular prism, but it may be in a different position.

Lesson 18:   Determining Surface Area of Three-Dimensional Figures

EUREKA MATH

5. Examine the figure below.

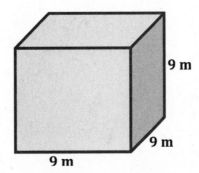

a. What is the most specific name of the three-dimensional shape?

   Cube

b. Write two different expressions for the surface area.

   $(9 \text{ m} \times 9 \text{ m}) + (9 \text{ m} \times 9 \text{ m}) + (9 \text{ m} \times 9 \text{ m}) + (9 \text{ m} \times 9 \text{ m}) + (9 \text{ m} \times 9 \text{ m}) + (9 \text{ m} \times 9 \text{ m})$

   $6(9 \text{ m})^2$

   > I know that a cube is a unique rectangular prism because it has 6 identical faces. I can use the formula for surface area of a rectangular prism or the formula $SA = 6s^2$, which is more efficient for a cube.

c. Explain how these two expressions are equivalent.

   The two expressions are equivalent because the first expression shows the sum of the areas of each face. The second expression is a more compact expression because each face has the same area and there are 6 groups of 9 m^2.

Calculate the surface area of each figure below. Figures are not drawn to scale.

1.

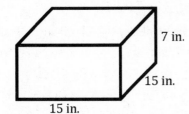

7 in.

15 in.

15 in.

2.

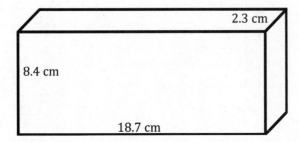

2.3 cm

8.4 cm

18.7 cm

3.

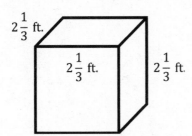

$2\frac{1}{3}$ ft.

$2\frac{1}{3}$ ft.

$2\frac{1}{3}$ ft.

4.

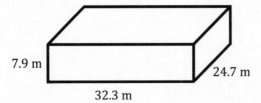

7.9 m

24.7 m

32.3 m

5. Write a numerical expression to show how to calculate the surface area of the rectangular prism. Explain each part of the expression.

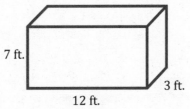

7 ft.

3 ft.

12 ft.

EUREKA MATH

6. When Louie was calculating the surface area for Problem 4, he identified the following:

    length = 24.7 m, width = 32.3 m, and height = 7.9 m.

    However, when Rocko was calculating the surface area for the same problem, he identified the following:

    length = 32.3 m, width = 24.7 m, and height = 7.9 m.

    Would Louie and Rocko get the same answer? Why or why not?

7. Examine the figure below.

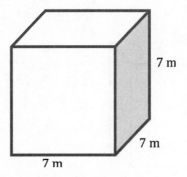

7 m

7 m

7 m

    a. What is the most specific name of the three-dimensional shape?

    b. Write two different expressions for the surface area.

    c. Explain how these two expressions are equivalent.

EUREKA
MATH

Opening Exercise

A box needs to be painted. How many square inches need to be painted to cover the entire surface of the box?

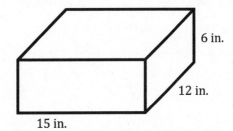

6 in.

12 in.

15 in.

A juice box is 4 in. tall, 1 in. wide, and 2 in. long. How much juice fits inside the juice box?

How did you decide how to solve each problem?

Discussion

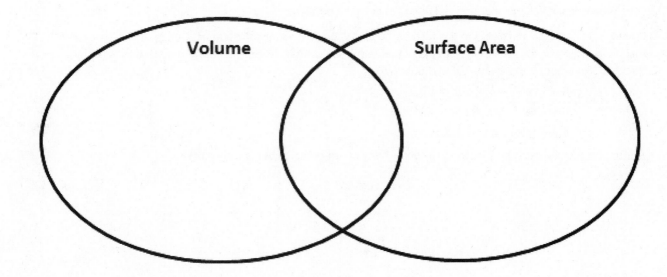

Volume   Surface Area

Example 1

Vincent put logs in the shape of a rectangular prism outside his house. However, it is supposed to snow, and Vincent wants to buy a cover so the logs stay dry. If the pile of logs creates a rectangular prism with these measurements:

33 cm long, 12 cm wide, and 48 cm high,

what is the minimum amount of material needed to cover the pile of logs?

Exercises

Use your knowledge of volume and surface area to answer each problem.

1. Quincy Place wants to add a pool to the neighborhood. When determining the budget, Quincy Place determined that it would also be able to install a baby pool that required less than 15 cubic feet of water. Quincy Place has three different models of a baby pool to choose from.

    Choice One: 5 ft. × 5 ft. × 1 ft.
    Choice Two: 4 ft. × 3 ft. × 1 ft.
    Choice Three: 4 ft. × 2 ft. × 2 ft.

    Which of these choices is best for the baby pool? Why are the others not good choices?

2. A packaging firm has been hired to create a box for baby blocks. The firm was hired because it could save money by creating a box using the least amount of material. The packaging firm knows that the volume of the box must be 18 cm^3.

    a. What are possible dimensions for the box if the volume must be exactly 18 cm^3?

    b. Which set of dimensions should the packaging firm choose in order to use the least amount of material? Explain.

3. A gift has the dimensions of $50 \text{ cm} \times 35 \text{ cm} \times 5 \text{ cm}$. You have wrapping paper with dimensions of $75 \text{ cm} \times 60 \text{ cm}$. Do you have enough wrapping paper to wrap the gift? Why or why not?

4. Tony bought a flat-rate box from the post office to send a gift to his mother for Mother's Day. The dimensions of the medium-size box are $14 \text{ inches} \times 12 \text{ inches} \times 3.5 \text{ inches}$. What is the volume of the largest gift he can send to his mother?

              Lesson 19:   Surface Area and Volume in the Real World   **237**

5. A cereal company wants to change the shape of its cereal box in order to attract the attention of shoppers. The original cereal box has dimensions of 8 inches × 3 inches × 11 inches. The new box the cereal company is thinking of would have dimensions of 10 inches × 10 inches × 3 inches.

    a. Which box holds more cereal?

    b. Which box requires more material to make?

6. Cinema theaters created a new popcorn box in the shape of a rectangular prism. The new popcorn box has a length of 6 inches, a width of 3.5 inches, and a height of 3.5 inches but does not include a lid.

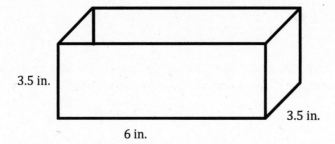

    a. How much material is needed to create the box?

    b. How much popcorn does the box hold?

EUREKA MATH®

Name _____   Date _____

Solve the word problem below.

Kelly has a rectangular fish aquarium that measures 18 inches long, 8 inches wide, and 12 inches tall.

    a.   What is the maximum amount of water the aquarium can hold?

    b.   If Kelly wanted to put a protective covering on the four glass walls of the aquarium, how big does the cover have to be?

1. Samuel built a small wooden box to hold nails. Each side of the box measures 7 inches.

   a. How many square inches of wood did he use to build the box?

   Surface Area of the Box: $SA = 6(7 \text{ in.})^2 = 6(49 \text{ in}^2) = 294 \text{ in}^2$

   Samuel used 294 square inches of wood to build the box.

   > I need to determine the surface area because it measures the total area of the surface of a figure.

   b. How many cubic inches of nails does the box hold?

   Volume of the Box: $V = 7 \text{ in.} \times 7 \text{ in.} \times 7 \text{ in.} = 343 \text{ in}^3$

   The box holds 343 cubic inches of nails.

   > The volume of a cube measures the space inside a three dimensional figure. To calculate the volume of a cube, I multiply the length, width, and height, or $V = s^3$ since the dimensions are the same.

2. A company that manufactures containers wants to know how many different containers it can make if the dimensions must be whole numbers and each container has a volume of 18 cubic centimeters.

   a. List all the possible whole number dimensions for the box.

   Choice One: $1 \text{ cm} \times 1 \text{ cm} \times 18 \text{ cm}$

   Choice Two: $1 \text{ cm} \times 2 \text{ cm} \times 9 \text{ cm}$

   Choice Three: $1 \text{ cm} \times 3 \text{ cm} \times 6 \text{ cm}$

   Choice Four: $2 \text{ cm} \times 3 \text{ cm} \times 3 \text{ cm}$

   > I know the dimensions of each container are factors of 18.

b. Which possibility requires the least amount of material to make?

Choice One:

$SA = 2(1 \text{ cm})(1 \text{ cm}) + 2(1 \text{ cm})(18 \text{ cm}) + 2(1 \text{ cm})(18 \text{ cm})$

$SA = 2 \text{ cm}^2 + 36 \text{ cm}^2 + 36 \text{ cm}^2$

$SA = 74 \text{ cm}^2$

> I need to calculate the surface area of each container to determine the one that requires the least amount of material.

Choice Two:

$SA = 2(1 \text{ cm})(2 \text{ cm}) + 2(1 \text{ cm})(9 \text{ cm}) + 2(2 \text{ cm})(9 \text{ cm})$

$SA = 4 \text{ cm}^2 + 18 \text{ cm}^2 + 36 \text{ cm}^2$

$SA = 58 \text{ cm}^2$

Choice Three:

$SA = 2(1 \text{ cm})(3 \text{ cm}) + 2(1 \text{ cm})(6 \text{ cm}) + 2(3 \text{ cm})(6 \text{ cm})$

$SA = 6 \text{ cm}^2 + 12 \text{ cm}^2 + 36 \text{ cm}^2$

$SA = 54 \text{ cm}^2$

Choice Four:

$SA = 2(2 \text{ cm})(3 \text{ cm}) + 2(2 \text{ cm})(3 \text{ cm}) + 2(3 \text{ cm})(3 \text{ cm})$

$SA = 12 \text{ cm}^2 + 12 \text{ cm}^2 + 18 \text{ cm}^2$

$SA = 42 \text{ cm}^2$

Choice Four requires the least amount of material because it has the smallest surface area.

c. Which box would you recommend the company use? Why?

I would recommend the box with dimensions of $2 \text{ cm} \times 3 \text{ cm} \times 3 \text{ cm}$ *(Choice Four) because it requires the least amount of material to make, which means it costs the company the least amount of money to make.*

EUREKA MATH

3. Auntie Math, Co. has two different boxes for Auntie Math Cereal. The large box is 7.5 inches wide, 8 inches high, and 3 inches deep. The small box is 4 inches wide, 11 inches high, and 1.5 inches deep.

a. How much more cardboard is needed to make the large box than the small box?

Surface Area of the Large Box:

$$2(7.5 \text{ in.})(8 \text{ in.}) + 2(7.5 \text{ in.})(3 \text{ in.}) + 2(8 \text{ in.})(3 \text{ in.})$$

$$120 \text{ in}^2 + 45 \text{ in}^2 + 48 \text{ in}^2$$

$$213 \text{ in}^2$$

> Before I can answer the question, I need to calculate the surface area for each box.

Surface Area of the Small Box:

$$2(4 \text{ in.})(11 \text{ in.}) + 2(4 \text{ in.})(1.5 \text{ in.}) + 2(11 \text{ in.})(1.5 \text{ in.})$$

$$88 \text{ in}^2 + 12 \text{ in}^2 + 33 \text{ in}^2$$

$$133 \text{ in}^2$$

Difference:

$$213 \text{ in}^2 - 133 \text{ in}^2$$

$$80 \text{ in}^2$$

The large box requires 80 square inches more material than the small box.

b. How much more cereal does the large box hold than the small box?

Volume of the Large Box:

$$7.5 \text{ in.} \times 8 \text{ in.} \times 3 \text{ in.}$$

$$180 \text{ in}^3$$

> Before answering the question, I need to calculate the volume of each box.

Volume of the Small Box:

$$4 \text{ in.} \times 11 \text{ in.} \times 1.5 \text{ in.}$$

$$66 \text{ in}^3$$

Difference:

$$180 \text{ in}^3 - 66 \text{ in}^3$$

$$114 \text{ in}^3$$

The large box holds 114 cubic inches more cereal than the small box.

4. A swimming pool is 9 meters long, 5 meters wide, and 2 meters deep. The water-resistant paint needed for the pool costs $5 per square meter. How much will it cost to paint the pool?

   a. How many faces of the pool do you have to paint?

   You have to point 5 faces. ⎯⎯⎯⎯⎯⎯⎯⎯⎯⎯⎯ | I do not have to paint the top of the pool. |

   b. How much paint (in square meters) do you need to paint the pool?

   Surface area of all six faces:

   $$2(9 \text{ m} \times 5 \text{ m}) + 2(9 \text{ m} \times 2 \text{ m}) + 2(5 \text{ m} \times 2 \text{ m})$$

   $$90 \text{ m}^2 + 36 \text{ m}^2 + 20 \text{ m}^2$$

   $$146 \text{ m}^2$$

   | I can also use the expression $9 \text{ m} \times 5 \text{ m} + 2(9 \text{ m} \times 2 \text{ m}) + 2(5 \text{ m} \times 2 \text{ m})$ since there is only ONE face with the dimensions $9 \text{ m} \times 5 \text{ m}$. |

   Area of Top of Pool:

   $$9 \text{ m} \times 5 \text{ m}$$

   $$45 \text{ m}^2$$

   Total Paint Needed:

   $$146 \text{ m}^2 - 45 \text{ m}^2$$

   $$101 \text{ m}^2$$

   c. How much will it cost to paint the pool?

   $$101 \times 5 = 505$$

   It will cost $505 to paint the pool.

5. The volume of Box X subtracted from the volume of Box Y is 16.2 cubic centimeters. Box X has a volume of 5.63 cubic centimeters.

   a. Let Y be the volume of Box Y in cubic centimeters. Write an equation that could be used to determine the volume of Box Y.

   $$Y - 5.63 \text{ cm}^3 = 16.2 \text{ cm}^3$$

   | Let X be the volume of box X. I know $Y - X$ is 16.2 cm^3, so I can substitute the volume of Box X in for X. |

Lesson 19:   Surface Area and Volume in the Real World

EUREKA MATH

b. Solve the equation to determine the volume of Box Y.

$$Y - 5.63 \text{ cm}^3 = 16.2 \text{ cm}^3$$
$$Y - 5.63 \text{ cm}^3 + 5.63 \text{ cm}^3 = 16.2 \text{ cm}^3 + 5.63 \text{ cm}^3$$
$$Y = 21.83 \text{ cm}^3$$

> I have a lot of experience with isolating the variable in an equation.

c. The volume of Box Y is one-tenth the volume of another box, Box Z. Let Z represent the volume of Box Z. Write an equation that could be used to determine the volume of Box Z, using the result from part (b).

$$21.83 \text{ cm}^3 = \frac{1}{10}Z$$

d. Solve the equation to determine the volume of Box Z.

$$21.83 \text{ cm}^3 = \frac{1}{10}Z$$
$$21.83 \text{ cm}^3 \div \frac{1}{10} = \frac{1}{10}Z \div \frac{1}{10}$$
$$218.3 \text{ cm}^3 = Z$$

Solve each problem below.

1. Dante built a wooden, cubic toy box for his son. Each side of the box measures 2 feet.

    a. How many square feet of wood did he use to build the box?

    b. How many cubic feet of toys will the box hold?

2. A company that manufactures gift boxes wants to know how many different-sized boxes having a volume of 50 cubic centimeters it can make if the dimensions must be whole centimeters.

    a. List all the possible whole number dimensions for the box.

    b. Which possibility requires the least amount of material to make?

    c. Which box would you recommend the company use? Why?

3. A rectangular box of rice is shown below. What is the greatest amount of rice, in cubic inches, that the box can hold?

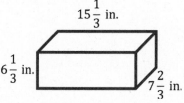

$15\frac{1}{3}$ in.

$6\frac{1}{3}$ in.

$7\frac{2}{3}$ in.

4. The Mars Cereal Company has two different cereal boxes for Mars Cereal. The large box is 8 inches wide, 11 inches high, and 3 inches deep. The small box is 6 inches wide, 10 inches high, and 2.5 inches deep.

    a. How much more cardboard is needed to make the large box than the small box?

    b. How much more cereal does the large box hold than the small box?

5. A swimming pool is 8 meters long, 6 meters wide, and 2 meters deep. The water-resistant paint needed for the pool costs $6 per square meter. How much will it cost to paint the pool?

    a. How many faces of the pool do you have to paint?

    b. How much paint (in square meters) do you need to paint the pool?

    c. How much will it cost to paint the pool?

6. Sam is in charge of filling a rectangular hole with cement. The hole is 9 feet long, 3 feet wide, and 2 feet deep. How much cement will Sam need?

7. The volume of Box D subtracted from the volume of Box C is 23.14 cubic centimeters. Box D has a volume of 10.115 cubic centimeters.

   a. Let C be the volume of Box C in cubic centimeters. Write an equation that could be used to determine the volume of Box C.

   b. Solve the equation to determine the volume of Box C.

   c. The volume of Box C is one-tenth the volume another box, Box E. Let E represent the volume of Box E in cubic centimeters. Write an equation that could be used to determine the volume of Box E, using the result from part (b).

   d. Solve the equation to determine the volume of Box E.

© 2019 Great Minds®. eureka-math.org

EUREKA
MATH

Opening Exercise

Determine the volume of this aquarium.

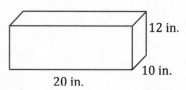

Mathematical Modeling Exercise: Using Ratios and Unit Rate to Determine Volume

For his environmental science project, Jamie is creating habitats for various wildlife including fish, aquatic turtles, and aquatic frogs. For each of these habitats, he uses a standard aquarium with length, width, and height dimensions measured in inches, identical to the aquarium mentioned in the Opening Exercise. To begin his project, Jamie needs to determine the volume, or cubic inches, of water that can fill the aquarium.

Use the table below to determine the unit rate of gallons/cubic inches.

Gallons	Cubic Inches
1	
2	462
3	693
4	924
5	1,155

Determine the volume of the aquarium.

Exercise 1

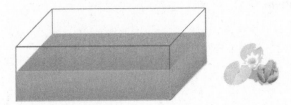

a.   Determine the volume of the tank when filled with 7 gallons of water.

b.   Work with your group to determine the height of the water when Jamie places 7 gallons of water in the aquarium.

Exercise 2

a.   Use the table from Example 1 to determine the volume of the aquarium when Jamie pours 3 gallons of water into the tank.

b.   Use the volume formula to determine the missing height dimension.

Exercise 3

a. Using the table of values below, determine the unit rate of liters to gallon.

Gallons	Liters
1	
2	7.57
4	15.14

b. Using this conversion, determine the number of liters needed to fill the 10-gallon tank.

c. The ratio of the number of centimeters to the number of inches is 2.54: 1. What is the unit rate?

d. Using this information, complete the table to convert the heights of the water in inches to the heights of the water in centimeters Jamie will need for his project at home.

Height (in inches)	Convert to Centimeters	Height (in centimeters)
1	$2.54 \dfrac{\text{centimeters}}{\text{inch}} \times 1 \text{ inch}$	2.54
3.465		
8.085		
11.55		

Exercise 4

a. Determine the amount of plastic film the manufacturer uses to cover the aquarium faces. Draw a sketch of the aquarium to assist in your calculations. Remember that the actual height of the aquarium is 12 inches.

b. We do not include the measurement of the top of the aquarium since it is open without glass and does not need to be covered with film. Determine the area of the top of the aquarium, and find the amount of film the manufacturer uses to cover only the sides, front, back, and bottom.

c. Since Jamie needs three aquariums, determine the total surface area of the three aquariums.

Name _____   Date _____

What did you learn today? Describe at least one situation in real life that would draw on the skills you used today.

1. Calculate the area of the figure below.

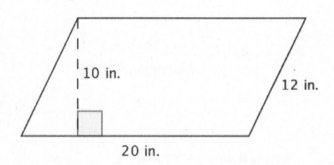

I can use the formula for area of a parallelogram.

$$A = bh$$
$$A = (20 \text{ in.})(10 \text{ in.})$$
$$A = 200 \text{ in}^2$$

2. Calculate the area of the figure below.

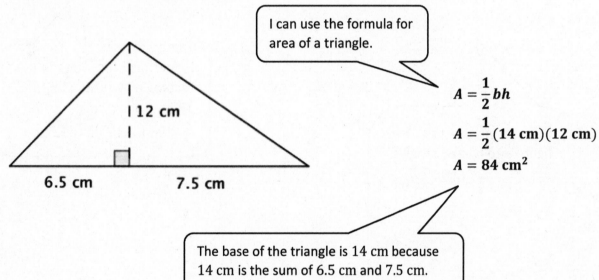

I can use the formula for area of a triangle.

$$A = \frac{1}{2}bh$$
$$A = \frac{1}{2}(14 \text{ cm})(12 \text{ cm})$$
$$A = 84 \text{ cm}^2$$

The base of the triangle is 14 cm because 14 cm is the sum of 6.5 cm and 7.5 cm.

3. Calculate the area of the figure below.

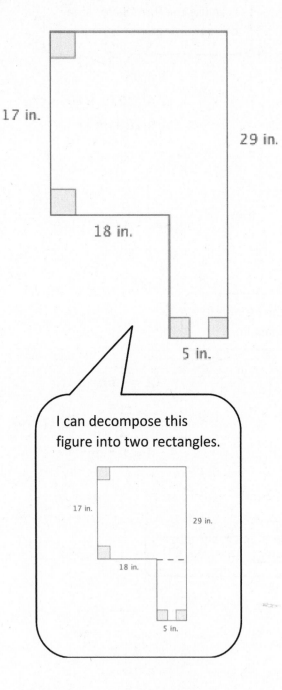

Area of top rectangle:

$$A = lw$$
$$A = (23 \text{ in.})(17 \text{ in.})$$
$$A = 391 \text{ in}^2$$

Area of bottom rectangle:

$$A = lw$$
$$A = (5 \text{ in.})(12 \text{ in.})$$
$$A = 60 \text{ in}^2$$

I determine the width of this rectangle by subtracting 17 in. from 29 in.

29 in. − 17 in. = 12 in.

I can decompose this figure into two rectangles.

EUREKA
MATH®

4. Complete the table using the diagram on the coordinate plane to find the distance between the two points on each segment.

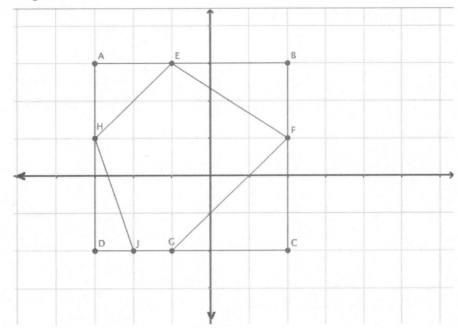

Line Segment	Point	Point	Distance	Proof
$\overline{AB}$	$(-3,3)$	$(2,3)$	5	$\lvert-3\rvert+\lvert2\rvert=5$
$\overline{BF}$	$(2,3)$	$(2,1)$	2	$\lvert3\rvert-\lvert1\rvert=2$
$\overline{CG}$	$(2,-2)$	$(-1,-2)$	3	$\lvert2\rvert+\lvert-1\rvert=3$
$\overline{DH}$	$(-3,-2)$	$(-3,1)$	3	$\lvert-2\rvert+\lvert1\rvert=3$
$\overline{HA}$	$(-3,1)$	$(-3,3)$	2	$\lvert3\rvert-\lvert1\rvert=2$
$\overline{AD}$	$(-3,3)$	$(-3,-2)$	5	$\lvert3\rvert+\lvert-2\rvert=5$

I find the location of each point in the line segment on the coordinate plane and record the coordinates of each point. If the points are on the same side of 0, I subtract the absolute values of the different coordinate to find the distance between the two points. If the points are on opposite sides of 0, I add the absolute values of the coordinates to find the distance between the two points.

5. Plot the points below, and draw the shape. Then, determine the area of the polygon.

$A(-3, 3), B(2, -1), C(0, -2)$

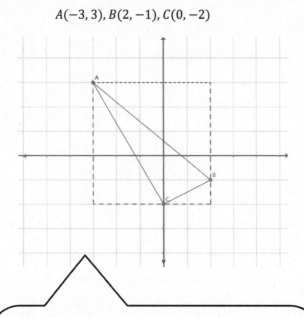

Area of rectangle:

$$\text{Area} = lw$$
$$\text{Area} = (5 \text{ units})(5 \text{ units})$$
$$\text{Area} = 25 \text{ units}^2$$

Area on left:

$$\text{Area} = \frac{1}{2}bh$$
$$\text{Area} = \frac{1}{2}(3 \text{ units})(5 \text{ units})$$
$$\text{Area} = 7.5 \text{ units}^2$$

Area on top:

$$\text{Area} = \frac{1}{2}bh$$
$$\text{Area} = \frac{1}{2}(5 \text{ units})(4 \text{ units})$$
$$\text{Area} = 10 \text{ units}^2$$

Area on right:

$$\text{Area} = \frac{1}{2}bh$$
$$\text{Area} = \frac{1}{2}(2 \text{ units})(1 \text{ units})$$
$$\text{Area} = 1 \text{ unit}^2$$

I can plot each point on the coordinate plane. To find the area of the polygon, I can draw a rectangle around the figure and find the area of the sections around the polygon. Since each of the surrounding sections is a triangle, I use the formula for area of a triangle to find the area of each section. Then, to find the area of the polygon, I can subtract the area of each section

Area of the Polygon $= 25 \text{ units}^2 - 7.5 \text{ units}^2 - 10 \text{ units}^2 - 1 \text{ unit}^2$

Area of the Polygon $= 6.5 \text{ units}^2$

EUREKA MATH

6. Determine the volume of the figure.

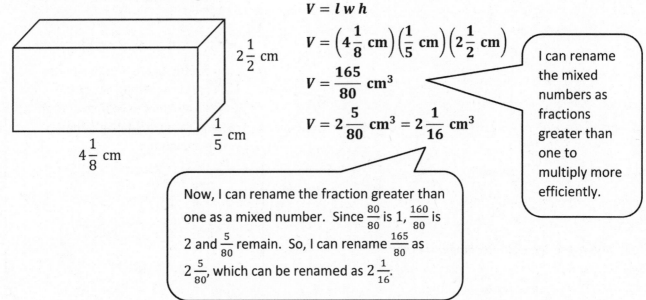

$V = l\,w\,h$

$V = \left(4\dfrac{1}{8}\text{ cm}\right)\left(\dfrac{1}{5}\text{ cm}\right)\left(2\dfrac{1}{2}\text{ cm}\right)$

$V = \dfrac{165}{80}\text{ cm}^3$

$V = 2\dfrac{5}{80}\text{ cm}^3 = 2\dfrac{1}{16}\text{ cm}^3$

I can rename the mixed numbers as fractions greater than one to multiply more efficiently.

Now, I can rename the fraction greater than one as a mixed number. Since $\dfrac{80}{80}$ is 1, $\dfrac{160}{80}$ is 2 and $\dfrac{5}{80}$ remain. So, I can rename $\dfrac{165}{80}$ as $2\dfrac{5}{80}$, which can be renamed as $2\dfrac{1}{16}$.

7. Give at least three more expressions that could be used to determine the volume of the figure in Problem 6.

Answers may vary. Some examples include the following.

$\left(\dfrac{1}{5}\text{ cm}\right)\left(2\dfrac{1}{2}\text{ cm}\right)\left(4\dfrac{1}{8}\text{ cm}\right)$

$\left(\dfrac{33}{40}\text{ cm}^2\right)\left(2\dfrac{1}{2}\text{ cm}\right)$

$\left(\dfrac{5}{10}\text{ cm}^2\right)\left(4\dfrac{1}{8}\text{ cm}\right)$

In this expression, I can rearrange the length, width, and height and then multiply.

Here, I can calculate the area of the base ($l \times w$) and then multiply by the height.

Because of the commutative property, I can multiply the width by the height ($w \times h$) and then multiply the product by the length.

8. Determine the volume of the irregular figure.

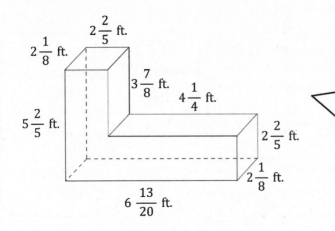

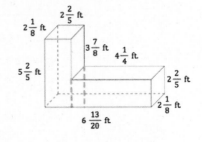

I can cut the figure into two parts. Each part is a rectangular prism. Then, I can find the volume of each rectangular prism and add the volumes together.

Volume of the right Rectangular Prism:

$$V = l\,w\,h$$

$$V = \left(4\frac{1}{4}\ \text{ft.}\right)\left(2\frac{2}{5}\ \text{ft.}\right)\left(2\frac{1}{8}\ \text{ft.}\right)$$

$$V = \frac{3468}{160}\ \text{ft}^3$$

$$V = 21\frac{108}{160}\ \text{ft}^3 = 21\frac{27}{40}\ \text{ft}^3$$

Volume of the left Rectangular Prism:

$$V = l\,w\,h$$

$$V = \left(5\frac{2}{5}\ \text{ft.}\right)\left(2\frac{1}{8}\ \text{ft.}\right)\left(2\frac{2}{5}\ \text{ft.}\right)$$

$$V = \frac{5,508}{200}\ \text{ft}^3$$

$$V = 27\frac{108}{200}\ \text{ft}^3 = 27\frac{27}{50}\ \text{ft}^3$$

$$\textbf{Total Volume} = 21\frac{27}{40}\ \text{ft}^3 + 27\frac{27}{50}\ \text{ft}^3 = 49\frac{43}{200}\ \text{ft}^3$$

EUREKA MATH

This Problem Set is a culmination of skills learned in this module. Note that the figures are not drawn to scale.

1. Calculate the area of the figure below.

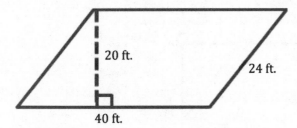

2. Calculate the area of the figure below.

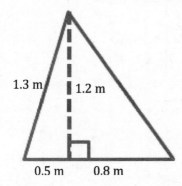

3. Calculate the area of the figure below.

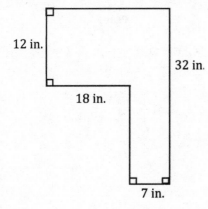

4. Complete the table using the diagram on the coordinate plane.

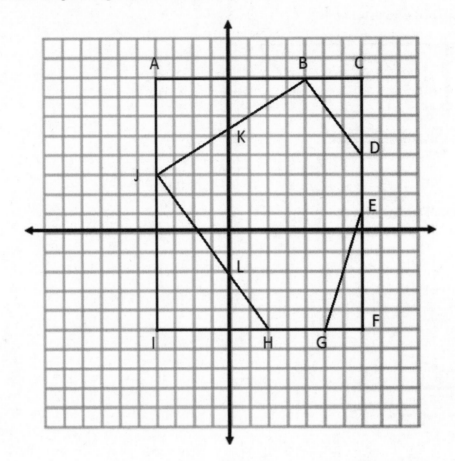

Line Segment	Point	Point	Distance	Proof
$\overline{AB}$				
$\overline{CE}$				
$\overline{GI}$				
$\overline{HI}$				
$\overline{IJ}$				
$\overline{AI}$				
$\overline{AJ}$				

Lesson 19a: Applying Surface Area and Volume to Aquariums

EUREKA
MATH®

5. Plot the points below, and draw the shape. Then, determine the area of the polygon.

$A(-3, 5), B(4, 3), C(0, -5)$

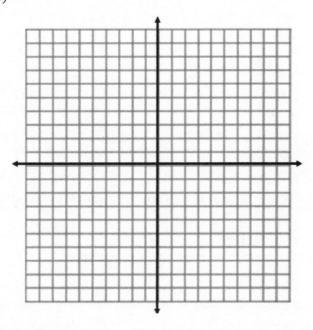

6. Determine the volume of the figure.

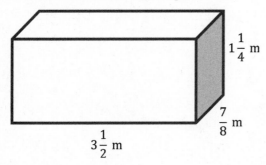

$1\frac{1}{4}$ m

$\frac{7}{8}$ m

$3\frac{1}{2}$ m

7. Give at least three more expressions that could be used to determine the volume of the figure in Problem 6.

8. Determine the volume of the irregular figure.

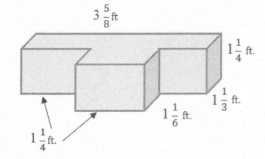

$3\frac{5}{8}$ ft

$1\frac{1}{4}$ ft.

$1\frac{1}{3}$ ft.

$1\frac{1}{6}$ ft.

$1\frac{1}{4}$ ft.

9. Draw and label a net for the following figure. Then, use the net to determine the surface area of the figure.

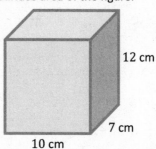

12 cm

7 cm

10 cm

10. Determine the surface area of the figure in Problem 9 using the formula $SA = 2lw + 2lh + 2wh$. Then, compare your answer to the solution in Problem 9.

11. A parallelogram has a base of 4.5 cm and an area of 9.495 cm². Tania wrote the equation $4.5x = 9.495$ to represent this situation.

   a. Explain what x represents in the equation.

   b. Solve the equation for x and determine the height of the parallelogram.

12. Triangle A has an area equal to one-third the area of Triangle B. Triangle A has an area of $3\frac{1}{2}$ square meters.

   a. Gerard wrote the equation $\frac{B}{3} = 3\frac{1}{2}$. Explain what B represents in the equation.

   b. Determine the area of Triangle B.

EUREKA
MATH

Credits

Great Minds® has made every effort to obtain permission for the reprinting of all copyrighted material. If any owner of copyrighted material is not acknowledged herein, please contact Great Minds for proper acknowledgment in all future editions and reprints of this module.

MANAGING LIBRARY EMPLOYEES

A How-To-Do-It Manual®

MARY J. STANLEY

HOW-TO-DO-IT MANUALS®

NUMBER 161

NEAL-SCHUMAN PUBLISHERS, INC.
New York   London

Published by Neal-Schuman Publishers, Inc.
100 William Street, Suite 2004
New York, NY 10038
http://www.neal-schuman.com

Printed and bound in the United States of America.

The paper used in this publication meets the minimum requirements of American National Standard for Information Sciences—Permanence of Paper for Printed Library Materials, ANSI Z39.48-1992.

Library of Congress Cataloging-in-Publication Data

Stanley, Mary J.
   Managing library employees : a how-to-do-it manual / Mary J. Stanley.
        p. cm. — (How-to-do-it manuals ; no. 161)
   Includes bibliographical references and index.
   ISBN 978-1-55570-628-9 (alk. paper)
   1. Library personnel management—Handbooks, manuals, etc. I. Title.
Z682.S76 2008
023'.9—dc22
                                                        2007051961